AF411417

# Large-Scale Urban Parks on Post-Industrial Sites in Contemporary Urban Landscape Conceptions

# Large-Scale Urban Parks on Post-Industrial Sites in Contemporary Urban Landscape Conceptions

Mengyixin Li

Basel • Beijing • Wuhan • Barcelona • Belgrade • Novi Sad • Cluj • Manchester

*Author*
Mengyixin Li
Beijing University of Civil
Engineering and Architecture
Beijing, China

*Editorial Office*
MDPI
St. Alban-Anlage 66
4052 Basel, Switzerland

For citation purposes, cite as indicated below:

Lastname, Firstname, Firstname Lastname, and Firstname Lastname. Year. *Book Title*. Series Title (optional). Basel: MDPI, Page Range.

**ISBN 978-3-0365-5560-7 (Hbk)**
**ISBN 978-3-0365-5559-1 (PDF)**
**doi.org/10.3390/books978-3-0365-5559-1**

Cover image courtesy of Danzi Wu.
Funding information: Beijing Project, Beijing University of Civil Engineering and Architecture.

# Contents

# List of Tables and Figures

Figure 37. Berlin Tempelhofer Feld, McGregor Coxall, 2010. Source: Photo by author.

Figure 38. Track Wilderness in Berlin's Park am Gleisdreieck, Atelier LOIDL, 2011. Source: Photo by author.

Figure 39. German cultural landscape revealing the harmonious relationship between people and land. Source: Photo by author.

Figure 40. The IBA Fürst-Pückler-Land project of Geopark Muskau Coal Crescent. Source: Photo by author.

Figure 41. Renaturated former railways in Park Duisburg-Nord. Source: Photo by ©Luca Maria Francesco Fabris; used with permission.

Figure 42. The urban wildness experienced by visitors in the mining region of Lusatia through IBA Fürst-Pückler-Land. Source: Photo by author.

Figure 43. The complex historical information segment concealed in the Berlin Wall. Source: Photo by author.

Figure 44. Landscape Park Duisburg-Nord interpreted as a kind of historical object with multiple layers of information. Source: Photo by author.

Figure 45. Shougang Industrial Heritage Park in Beijing, probably influenced by Peter Latz's philosophy in the process of reconsidering the transformation of post-industrial sites in China. Source: Photo by author.

Figure 46. Bürgpark Hafeninsel's existing layers of site information in Saarbrücken kept for the syntactical design, Latz + Partner, 1999. Source: Photo by ©Qi Huang 2016; used with permission.

Figure 47. The hilly terrain of Landscape Park Duisburg Nord organized through complex site elements with the contextualistic–structuralistic approach. Source: Photo courtesy of ©Latz + Partner; used with permission.

Figure 48. An impressive night-time experience on abandoned F60 conveyor bridge in IBA Fürst-Pückler-Land. Source: Figure by author.

Figure 49. Seventeen cities interconnected through a green corridor of Emscher Park, Gelsenkirchen. Source: Photo by ©Danzi Wu 2015; used with permission.

Figure 50. The relationships between Landscape Park Duisburg-Nord and its surroundings in the urban region. Source: Figure courtesy of ©Latz + Partner; used with permission. Figure 51. Building visual relations between Landscape Park Duisburg-Nord and its surroundings. Source: Figure courtesy of ©Latz + Partner; used with permission.

Figure 52. The abstract structure of Landscape Park Duisburg-Nord developed through overlay and connection of layers and structural elements. Source: Figure courtesy of ©Latz + Partner; used with permission.

Figure 53. The "water canal" designed for ecological restoration in Landscape Park Duisburg-Nord. Source: Figure courtesy of ©Latz + Partner; used with permission.

Figure 71. Master plan of Shougang Industrial Heritage Park, Tsinghua Urban Planning and Design Institute, 2016. Source: Figure courtesy of ©Tsinghua Urban Planning and Design Institute, Zhu Yufan Studio; used with permission.

Figure 72. The site complexity originating from the overlay of industrial remains, traditional gardening and huge modern elements. Source: Photo by ©Bin Li 2015 (top) and author (bottom); used with permission.

Figure 73. The shan-shui structure shaped by the park. Source: Photo by author.

Figure 74. Low-impact rainwater gardens of Shougang Industrial Heritage Park in the idea of sponge city. Source: Photo by author.

Table 1. Three large-scale urban park paradigms with essential cases in the study areas. Source: Author's compilation based on data from Weilacher 2008; Fabris and Li 2020; Li 2023.

Table 2. Three renewed urban models, their main features, and related specialists and tasks. Source: Table by author.

Table 3. Organic parks' new perspectives compared with conventional parks' stereotypical perspectives in terms of concept, positioning, role, and focus. Source: Author's compilation based on data from Cranz 1982; Corner 1999; Lister 2007; Marton 2010; Waldheim 2006; Czerniak 2001; Prominski 2010.

Table 4. Comparison of contemporary urban landscapes in two theoretical conceptions at three levels. Source: Author's compilation based on data from Latz 2008b; Corner 1997; Corner 1999.

Table 5. Five qualitative characteristics for the qualification of two large scale urban parks. Source: Table by author

Table 6. Relationships among critical rationalism approach, cultural embedding, core of shaping urban landscape and park identity for North America and Germany. Source: Table by author.

# Abbreviations

| | |
|---|---|
| AFCD | Agriculture, Fisheries, and Conservation Department. |
| BMU | Bundesministerium für Umwelt, Naturschutz und Nukleare Sicherheit. |
| BUGA05 | Bundesgartenschau 2005 (Federal Garden Show). |
| GSD | Harvard University Graduate School of Design. |
| IBA | Internationale Bauausstellung (International Building Exhibitions). |
| LAF | Landscape Architecture Foundation. |
| OMA | The Office for Metropolitan Architecture. |
| ISOCARP | International Society of City and Regional Planners. |
| LMRSB | Landeshauptstadt München Referat für Stadtplanung und Bauordnung. |
| NYCDPR | New York City Department of Parks & Recreation. |

# About the Author

Mengyixin Li received her Ph.D. in landscape architecture from the Technical University of Munich (TUM) in Germany, is an associate professor at the School of Architecture and Urban Planning, Beijing University of Civil Engineering and Architecture (BUCEA) in China, and is a member of the Territorial Landscape Committee, the Cultural Landscape Committee and the Landscape Theory and History Committee of the Chinese Society of Landscape Architecture, a member of TUM China Urban Landscape LAB at the Professorship of Landscape Architecture and Regional Open Space (LAREG), and the contributing editor of the journal *Landscape Architecture* in China. Currently, her research is focused on green open spaces, post-industrial landscapes and cultural landscapes.

# Foreword

Did you say large parks?

The word *park*, which maintains the same root in most parts of contemporary European languages, is relatively new. It was born in the Middle Ages and derived from the proto-German **barō*, which meant barrier. This barrier was a hunting enclosure, a privilege of the patrician families of the time. In northern Italy, we have many toponyms that recall the existence of *barcos* in areas that were once lush with woods and forests because hunting, understood as noble leisure, required an ideal environment, a perfect and uncontaminated place that proved that one had total control of nature. The *barcos* were the first large parks, and even if they were hunting reserves, they had many similarities with our contemporary parks, especially urban ones, which we frequent daily. Regardless of their appearance, current urban parks reflect a domesticated naturalness and the fundamental human need to experience a moment of diversion and entertainment. The urban public park in Europe developed in close coordination with industrialization when the overwhelming growth of cities began. The public gardens first and then the urban parks were open spaces that allowed an ever-increasing part of the population to freely spend their resting time, granted to workers of all classes, from the workforce to the bourgeoisie.

For this reason, public parks are also a quantifiable indicator of social policies and their evolution over time, so much so that they have become an indicator of healthiness in the rankings that evaluate our contemporary cities. Metropolitan conurbations, in the last fifty years, have been going through a new transformation, marked by the relocation of former industries and the technological–information transformation that has initiated the post-modern phase of 'non-work', understood as the end of hard work. Over the years, the closure of industrial plants, which started in North America and then spread to Europe and the Far East, has highlighted the problem of reactivating these abandoned areas, which have become integral parts of the city's urban fabric. A series of temporal synergies such as the birth of ecological awareness, the growth of policies for sustainability, the changing times of the city, the attention to health and the demands of citizens, increasingly attentive to environmental issues and the use of own time, have contributed to the excellent decision to transform many of the black holes resulting from deindustrialization into new urban parks, the large parks covered by this publication.

Mengyixin Li skillfully traces the history of large-scale parks by describing both the political strategies and the system of theories that, over time, have formed the basis of this process that has recently changed not only our cities and our metropolises but also the mentalities of our designers. The author, through the analysis of projects planned and built on abandoned industrial areas, allows us to understand this still-moving picture in all its peculiarities and facets, demonstrating how the theme of environmental redevelopment in urban areas is not one of the themes of contemporaneity, but *is* the theme.

Furthermore, this volume proves to be fundamental for the original plot that manages to weave between Western and Eastern specificities, specifically the Chinese ones, where environmental design combines with a millenary vision of balance with nature, which must be recovered and adapted to contemporary times.

Taking up the incipit of this introduction, Mengyixin Li leads us to understand and deepen the cultural and theoretical bases that underlie the new barcos: the new contemporary urban places where the topics of urban sustainability merge with landscape architecture and with the renewed need to reconnect with nature, even if tamed, during our leisure time as citizens.

There is a need to design new large parks without barriers and tailored to our large communities.

**Luca Maria Francesco Fabris**
**Politecnico di Milano—BUCEA Expert**
**Milan, Italy**

# Preface

The Anthropocene epoch refers to the time period in which human action has had a significant enough impact on Earth for a new geological era to begin. The collapse of biological habitats and global warming are two significant phenomena that have significantly changed the Earth's systems and are now epochal representations of the Anthropocene, attesting to the fact that humanity has now colonized every square mile of the planet.

Planning and design are two ways that people interact with the land in various cultures. However, we observe that it does not just alter the physical character and structure of regions and urban landscapes; we are also becoming more conscious of the drawbacks and potential side consequences of such interventions. While the landscape or nature should have the ability to evolve on its own throughout time, it appears that by planning and design, we can only influence a portion of the landscape or ecological processes that may be subject to numerous uncertainties, such as disease, earthquakes and floods. We are living in uncertain times, shaping our future in a transforming world, as noted in the latest UN Human Development Report 2021–2022.

Large-scale urban parks and abandoned industrial locations both show the effects of human activity on the city in this process. From a historical standpoint, the two appear to be at odds: while post-industrial sites are places that present themselves as abandoned, residual, chaotic, disorganized, and dirty, large-scale urban parks were originally created for the sanitation of urban environments, synonymous with beautification. The combination of these two is currently the subject of extensive research into the preservation and reuse of industrial remains as well as the regeneration, rehabilitation and redefinition of urban open spaces throughout the world. They demonstrate both landscape change and reflection in the present.

On the one hand, large-scale urban parks serve as a significant conduit for green open spaces and are a recurrent research topic of academic and popular interest, providing possibilities for individuals to interpret the landscape in their own unique ways. Large-scale urban parks have been developed with the input of many people. Large-scale urban parks may be subject to spatio-temporal variation, landscape theoretical advancement, natural succession, social appropriation, etc., as part of their planning and design, which constitute common tasks and challenges for landscape architects worldwide.

On the other side, mothballed industrial plants, contaminated brownfields, abandoned military bases, and landfills are now the standard post-industrial sites for large-scale urban parks. These kinds of sites generate "a brand-new park" (Kirkwood 2004) with site constraints (Czerniak and Hargreaves 2007) and other challenges with socio-economic structural transformation, especially in vulnerable peri-urban areas, because there are no other large expanses of land in urban centers. The cross-cultural research of large-scale urban parks on post-industrial sites is essential and in keeping with the times in light of the massive shift that urban realities are going through. According to Swiss landscape architect Günther Vogt (Grosch and Petrow 2021), one of the most challenging issues for our profession continues to be establishing a contemporary type of park.

In light of this, numerous studies highlight contemporary urban landscapes, post-industrial landscapes and sizeable urban parks primarily on reclaimed land in North America, Germany and China, respectively. The three regions are regarded as representative and more ideal for performing cross-cultural comparison and communication given the author's cultural background as a Chinese person, her educational experiences while studying abroad, and her global outlook. Furthermore, a range of studies not only demonstrate that iconic and typical post-industrial landscapes in urban areas have become a research topic of great interest to landscape architects throughout the world but also reflect that landscape architects in the regions have been major promoters and contributors to the theory and practice of post-industrial landscape as well as large-scale urban derelict land conversion. In the introduction, many key large parks and post-industrial urban projects are listed to underline the importance of the three study areas.

That said, few researchers—especially those with training in landscape architecture—combine critical urban landscape theories, conceptual frameworks, and practical experiences, analyze these elements on a par with case studies of large-scale urban parks, and then combine these findings. Given this, this book combines the transformation of industrial wastelands with specific large park models in the context of post-industrial landscape in the West and China, providing a more diverse research dimension at both the theoretical and practical levels, and expanding the critical perception and systematic understanding of urban parks and post-industrial landscapes.

In essence, the current study intends to clarify the critical theories that are now being created regarding urban landscapes and large-scale urban parks in relation to regional cultural contexts and to identify the key features that set developed urban landscapes apart from less developed ones. The stark contrast points to a deeply ingrained cultural "disposition", similar to a person's innate mental and moral makeup. The creative versus coherent cumulative understanding of landscapes between North America and Germany is one such example. Finding and comparing the differences between the concepts in terms of ideas and projects might help advance the research that will be conducted.

The potential application of the concept of large parks in the Chinese context is another reason for performing the current study. In terms of professional knowledge of the urban landscape and associated conceptual approaches from a critical perspective, Chinese researchers are more or less uninfluential. However, there are theoretical and practical ties between North America, Germany, and China for professionals in landscape architecture. For the benefit of Chinese landscape architects, this study tries to clarify the fundamental ideas behind urban landscape theories and projects in developed regions.

Consequently, this book about expansive urban parks on former industrial sites has been written after conducting a critical analysis of urban landscapes in North America, Germany, and China. The following is suggested as the research's main problem:

How should contemporary large-scale urban parks be viewed in light of urban spatial organization, society, and ecology within the context of culturally shifting conditions? Or, more specifically, how is the connection between expansive urban parks, urban nature, and contemporary cities being reimagined?

Based on the hypothesis, the research methodology of critical rationalism in the field of philosophy is applied, because its essence of falsifiability has been influential in Western planning culture since the late 1960s. In this book, critical rationalism provides a vital approach to critical thinking and a great impetus for the changing connotations of urban landscapes in the three regions. Specifically, two existing and crucial methodological approaches based on critical rationalism are discussed in the field of Western landscape architecture: James Corner's *critical thinking* for process-ecological methods and Peter Latz's *critical structuralism* for context-syntactical methods. Accordingly, there are three large park design paradigms: North American organic parks with a large-scale landscape architectural ideas for eco-oriented urban landscapes, German structuralistic parks with regional thinking for step-by-step landscape transformation, and Chinese shan-shui (mountain–water) parks with an idea of further exploring the texture of the cultural landscape. Conclusions about the dynamic contemporary and cultural contexts relating to urban spatial structure, society, and ecology, as well as potentials in the theoretical and practical advancements in international landscape design, are drawn from comparative examinations of these parks.

Furthermore, while Peter Latz's German structuralistic park model emphasizes structures based on cultural contextualization, James Corner's North American organic park model emphasizes designs from *cultural imagination*. Both concepts, which mainly emphasized the effects of regional cultural identities and ecological balancing, were created from the social uses and ecological function of large-scale urban parks. All theories and projects relating to complexity, diversity, sustainability, appropriation, and identity make use of these concepts.

The current study compares the large-scale urban park models' theoretical and practical circumstances. Theoretical concerns emphasize the urbanistic concepts of urban landscapes, *landscape urbanism* in North America, *careful renewal* or *critical reconstruction* of European cities, and regional landscape development in Germany. The practical considerations relate to specific design projects.

The landscape architectural park models and urbanistic theoretical frameworks in China are examined using the analytical results of the two models. In this context, similarities and differences between the existing Chinese urban landscapes of large-scale urban parks on post-industrial sites are examined. Thus, different sociocultural, ecological, and aesthetic changes might be influenced by international park models.

In this book, the frameworks of North American *landscape urbanism* and German *landscape structuralism*, as well as their two large-scale urban park design paradigms, are examined. They show striking parallels and differences between the two cultures' parks of landscapes (coherent vs. creative), landscape and ecology (representation vs. metaphor), and landscape and life (diversity vs. unpredictability). These analytical findings are used to reevaluate the third cultural model, Chinese shan-shui parks, and are characterized as cultural interpretations.

**Mengyixin Li**
**Beijing University of Civil Engineering and Architecture**
**Beijing, China**

# Acknowledgements

The book presents a summary of my academic research conducted on landscape architecture during the Professorship of Landscape Architecture and Regional Open Space at the Technical University of Munich in Germany as well as at the Beijing University of Civil Engineering and Architecture in China. I would like to express my sincere gratitude to Prof. Dr. Sören Schöbel-Rutschmann for offering strong support to my exploratory cross-cultural study and enlightening guidance in terms of European urban landscape theories and methodologies. I also much appreciate my beloved family for their unconditional love and encouragement, whether I was in Europe or China.

# 1. Introduction

"I would like to think that in the future the profession of landscape architecture will expand beyond its present confines and concern itself with making mobility orderly and beautiful. This would mean knowing a great deal about land, its uses, its values, and the political and economical and cultural forces affecting its distribution." (Jackson 1984)

Large-scale urban parks have been used as a concept for contemporary landscape planning and design. These parks are intrinsically tied to the development of contemporary cities, the various conceptions of dynamic urban landscapes, and the sustainable, cost-effective, and process-oriented transformation of post-industrial sites.

This book offers one of the first thorough analyses of contemporary large-scale parks on post-industrial sites in North America, Germany, and China in light of the importance of contemporary parks in the restoration, regeneration, and redevelopment of urban regions. This is done in an effort to reveal the essence, functions, characteristics, and potentials of parks on both a theoretical and practical level. We can discover the force of landscape at the urban level, as well as the visions, attitudes, and behaviors of contemporary park landscapes in various socio-cultural and ecological contexts, through an in-depth analysis of North American organic parks, German structuralistic parks, and Chinese shan-shui parks.

In this book, large-scale parks and urban landscapes are explored in the context of critical rationalism, which was proposed by Austrian-British philosopher and one of the twentieth century's most significant thinkers Karl Popper in 1957. This approach can offer illustrative guidelines for selecting theories to examine rather than hold fast to. This methodology is regarded as a scientific critical approach because a scientist, whether a theorist or investigator, presents statements or systems of claims and tests them step by step (Popper 1959). This is one way that the scientific examination of landscape theories related to contemporary parks and cities is connected to doubt, criticism, and denial. Based on this approach, diverse conceptions of urban landscapes are explored, and some essential parameters of large-scale parks are defined in terms of their size, vision, conception, transformation, planning and design approach, qualification and practice, through which the qualitative and quantitative analysis of different park paradigms are further manifested.

## 1.1. Diverse Conceptions of Cities and Urban Landscapes

Everything flows, and nothing is permanent. Cities and urban landscapes are consistent with this pattern. The constant evolution of contemporary cities with rapid growth or gradual shrinkage around the world not only brings a variety of opportunities for sustainable urban redevelopment and renewal but also great challenges in terms of changing the urban spatial structure, society and ecology.

Particularly, as the urban environment has shifted in the transition from an industrial to post-industrial society, many technical terms conceived by specialists for contemporary cities have emerged in the space-related fields (architecture, urbanism, landscape architecture, etc.) from the perspectives of urban structure (such as *Compact City, Porous City*), society (such as *Healthy City, Park City*) and ecology (such as *Sponge City, Resilient City*).

What do these ideas of contemporary cities imply for us? These terms essentially capture a variety of logical conclusions about the city in general, taking into account its complicated phenomena or realities, difficult issues that must be solved, people's critical thinking, and clever planning and design solutions. The cities from various cultural perspectives help us to describe how diverse and dynamic they are. They also help us to understand the relevance of conceptual cities in a world shaped by humans. Moreover, it is evident that our conceptions of cities, which are rich in complexity, are the beginning point for comprehending and visualizing urban landscapes.

Our conceptualisation of landscapes and parks at the urban level may also need to be considered in a dynamic way to adapt to the changes in our understanding of ideas such as the city, society, ecology, and aesthetics in contemporary landscape architecture, as long as we accept that cities, landscapes, and parks are integrated with each other and are all exposed to transformation in both reality and theory. Considering this aspect, landscape historians have demonstrated that landscapes flow with the times and is perpetually changing (Bucher 2013).

J. B. Jackson, an American expert on cultural landscapes, first noticed the trend in 1984 when he noticed how cities and the landscape were being altered. In his proposal to reimagine the *dynamic landscape* as "systems of man-made spaces on the surface of the earth," he has explored issues pertaining to landscape forms and identities (Jackson 1984). This definition suggests an expanded concept of *landscape*, reflecting "an open-minded, progressive, use-related landscape that was oriented toward modernity's dynamic transformations" (Truniger 2013), as opposed to *aesthetic landscape* or *scenic landscape*, the dominated and harmonious landscape with picturesque scenes in the tradition of Arcadia. A growing number of landscape architects in the West began to realize that the landscape concept had diverse meanings and contexts of usage (Bucher 2013). As a result, at the end of the twentieth century, the theory of landscape underwent a considerable transformation.

It Is important to clarify an overview of our research area of *urban landscape* in order to examine large-scale urban parks on post-industrial sites as important manifestations. This is done primarily on the basis of dynamic cities and landscapes from multiple perspectives.

Research interests were sparked in North America (USA, Canada), Germany, and other countries in the late twentieth century by newly developed theoretical analyses and conceptions of the contemporary *urban landscape* (Li 2023). There is not really a simple concept to describe the *urban landscape*. The conceptualizations of the two developed regions differ, which is what is responsible for the overall theoretical disagreements surrounding this concept. In North America and Europe, the urban disintegration issues were directly related to post-industrialization. Thus, a quest for alternative spatial structures — specifically, *urban landscape* — in many cultures was started. Theoretical analyses of *urban landscape* in North America and Europe from the 1970s to the 1980s were contributed to by J. B. Jackson and French philosopher and sociologist Henri Lefèbvre, respectively, according to the research.

By the 1990s, advanced *urban landscape* formulations were influenced by earlier works and strongly tied to the *critical thinking* and *critical structuralism* developed by American landscape architect and theorist James Corner and German landscape architect Peter Latz through the study of their theories and designs. As an interpretation of Peter Latz's published *structuralistic approach, Syntax of Landscape, critical structuralism* is employed. This book examines *landscape urbanism* and *landscape structuralism*, two opposing schools of thought in landscape architecture, in addition to the two philosophical stances in planning and design. With the passage of time and the help of theoretical assessments of conceptions in North America and Germany, a certain urban landscape"'s understanding evolves.

The analysis conceptualizes the contemporary *urban landscape* as a comprehensive yet multivalent concept intimately linked with urban society, urban structure, and urban nature based on these discussions in developed regions in recent decades. This concept not only allows for regionality and localization but also demonstrates that contrasted with omnipresent urbanization and globalization is a focus on regional cultural characteristics in planning and design. The technical term *urban landscape*, which is closely associated with contemporary cities and large-scale urban parks, is proposed in this book:

- in response to the transition to a post-industrial society, where urban remediation and renewal projects are developed to integrate complex site environments, public infrastructure, and urban everyday life;
- as a spatial concept in lock-step with a change in urban structure, the dissolution of the dominant urban organizational form in the process of massive urban growth, and the rise of suburbia;
- as a conceptual open structure offering diverse spatial forms to preserve urban nature in the face of ecological crisis and movement and to support and feed natural processes for the resilience of urban nature;
- as a positive term substituting for all other concepts, such as *Zwischenstadt (In-Between City), Edge City, Suburbania, Sprawl,* and *Periphery*;
- as a cultural construct within the scope of the cultural landscape full of evidence of human intervention, conception and exploration. Based on this, large-scale urban parks can serve as a concept for contemporary urban landscape planning in a cross-cultural discussion.

The aforementioned five levels suggest that in the process of making it easier to conceptualize expansive urban parks, theoretical frameworks for *urban landscape* need to evolve. These important sentences also discuss how contemporary urban landscapes have changed in terms of urban spatial organization, society, and ecology. The inclusion of the qualifier "urban" in the research to define terms such as landscape, society, structure, nature, life, and infrastructure was carried out to acknowledge the development of the wide idea of the *urban region* amongst twentieth-century European academics. The conceptualistion by German architect and urban planner Thomas Sieverts, who said that "the city is integrated with the landscape, and the old contrast between town and country has already substantially dissolved in favor of a city-landscape continuum," has had a significant influence on the term *urban region* (Sieverts 2003). Due to urban breakdown, the *urban region* specifically denotes an improvement in the spatial understanding of contemporary cities. The urban and the rural, cities and landscapes no longer maintain a condition of conflict in North America and Europe.

In addition, due to the understanding by landscape architects that landscape plays a crucial role in urban reconstruction and regeneration, conceptions of the *urban landscape* through *critical thinking* are changed in various ways. In other words, the landscape is becoming more and more of a protagonist in Western thinking. Regarding growing discourses on contemporary urban landscapes, American-Canadian architect and urbanist Charles Waldheim asserted that professional and critical categories have been validated to account for the revived interest in landscape found in the work of numerous architects, landscape architects, and urbanists over the past several years (Waldheim 2006). The contemporary trend in landscape design, as highlighted in the current study, is truly described by the "professional" and "critical" strands of Charles Waldheim.

In this book, Karl Popper's critical rationalism is credited with sparking the development of urban landscapes in the professional field of landscape architecture, which consists of a corpus of theoretical presumptions and exploratory methods. *Critical thinking* and *critical structuralism* are two kinds of critical research approaches that are mentioned. Parallel to them, researchers in North America and Germany developed critical theories and enhanced their understanding of contemporary urban landscapes. The current study focuses on these two critical conceptual frameworks, which are discussed in Chapter 3: North American *landscape urbanism* with an ecological or organic approach and German urban landscape with a *structuralistic approach*.

In the new understanding of "the dynamic nature of the material itself" (Berrizbeitia 2007), North American *landscape urbanism* with an ecological approach is process-orientated and views the "formation of space through process" or the process as the "principal generators" of space-making (Wall and Dring 2015). The American-Canadian theorist and activist Jane Jacobs has used the process-oriented approach since the 1960s, when she began to include biology as a factor in understanding the concept of contemporary cities. James Corner's conceptual premise of "a more organic, fluid urbanism" was particularly influential in how the contemporary city was depicted given the "professional" and "critical" views towards North American urban landscape at the turn of the twentieth century (Corner 2006). After going through this process, advocates of *landscape urbanism* employ an ecological approach.

The German *structuralistic approach*, however, is typically referred to as the foundation of scientific structuralism (Culler 2002). According to the definition of *Structuralism*, it is a theory that reconstructs systems of relationships through the use of culturally related signs rather than by examining isolated, physical objects in isolation (ibid.). Since the early twentieth century, this technique has been widely applied in numerous fields. In order to build a context-syntactical method, the current study elaborates on the *structuralistic approach*, which was first developed in structural linguistics and later applied to European architecture, German landscape architecture, and the interpretation of *critical structuralism*.

The use of the term *structure* was established among the structuralists in architecture in the wake of criticism of modernist functionalism, greatly influenced by the ideas of Swiss architectural theorist Arnulf Lüchinger and Dutch architect and structuralist Willem van Bodegraven. In the mid-1980s, structuralism became the focus of interest in landscape architecture when the ecology movement became increasingly strong and ways were sought to improve the ability to change, develop, and transform designed urban landscapes (Weilacher 2018). When German landscape architects accepted that a living landscape or city should be able to change its face without losing its face, they recognized that open structures are perfectly suited to developing growth-enabling integration that can change dynamically without losing their identity and inner cohesion (ibid.). Moreover, following French landscape architect Bernard Lassus, Swiss sociologist Lucius Burckhardt, and Peter Latz in Chapter 4, the *structuralistic approach* is connected to *minimal intervention* or the *smallest possible intervention* (Weilacher 2008). The *structuralistic approach* is used to deal with the examination of complex landscape systems with multiple structural layers and to cultivate diverse spaces for social appropriation in daily life.

In response to the aforementioned *urban landscape* with its five essential aspects, the term *Large-scale Urban Parks* is proposed in this book as a response. This is based on the evolving cities and *urban landscape* with diverse, critical conceptions boosting the development of contemporary park concepts. The subject of how we should envision contemporary, large-scale urban parks, particularly in post-industrial areas, is brought up in the meantime.

In actuality, various voices contribute to the development of the term "park" when it is highlighted at the urban level. From a diachronic perspective, there are *Modern Parks* and *Historic Parks*; in terms of nature conservation and use, there are *National Parks, Country Parks, Forest Parks*, and *Ecological Parks*; in terms of scale, there are *Large Parks* and *Pocket Parks*; and in terms of spatial integration and transformation, there are *Regional Parks and Post-industrial Landscape Parks*.

Is it equally feasible for contemporary urban parks to develop a vocabulary and conceptual expressions, similar to the numerous concepts of contemporary cities with various qualifiers in North America, Europe, and China. Large-scale urban parks will serve as a medium in this landscape so that we can gain insight into how urban park landscapes have changed to accommodate shifting cultural norms in terms of urban spatial organization, society, and ecology.

## 1.2. The Paradigm Shift of Large-Scale Urban Parks

Large-scale parks are frequently found in metropolitan regions. Examples include New York's Central Park (340 hectares), London's Hyde Park (140 hectares), Berlin's Tiergarten (210 hectares), Landscape Park Duisburg-Nord in Duisburg- Meiderich (230 hectares), Northern Milan Park (640 hectares), Beijing's Olympic Forest Park (680 hectares), etc. For example, London's Hyde Park is the largest of the royal parks connecting the English garden history with people's modern life, in which visitors can perceive historical landscape elements and newly designed places (such as the Diana Princess of Wales Memorial Fountain) for contemporary demands of attractive open green spaces, as seen in Figure 1. Over time, large-scale urban parks have developed into a rich source of landscape elements and site information, and they are an indispensable asset for cities in the past, present and future. Thus, more awareness and attention should be paid to the distinctive parks that serve as everyday landscapes for individuals in many countries.

**Figure 1.** Diana Princess of Wales Memorial Fountain in London, Gustafson Porter + Bowman, 2004. Source: Photo by author.

When we discuss them, the question "why do cities need large-scale parks?" may already be at the forefront of our minds. In fact, the same question came up at an international conference on "Large Parks in Large Cities" that was held in Stockholm in 2015. This symposium is regarded as a turning point in stressing the value of having nature adjacent to cities (Wirtén 2022).

From a historical standpoint, there has been a trend of urban sprawl that dates back to at least 1800 and is still present now. According to Richard Murray, an environmentalist and co-chair of the World Urban Parks Association's Large Urban Parks Committee, given this tendency in the majority of cities around the world, people constantly try to seize the chance to create resilient cities that are both compact and green through large urban parks (Murray 2022). In light of the significance of large-scale parks, American urban ecologist Richard T. T. Forman also demonstrates that "large parks are better than small parks" in terms of the essential advantages to cities, such as air conditioning, biodiversity, flood mitigation, and recreation (Forman 2022).

Since Frederick Law Olmsted's time, landscape architects all over the world have been thinking about what parks are, how they appear, and the functions they play in cities (Czerniak 2022) in order to fulfill the co-development of cities and large urban parks. He was a leading and well-known landscape architect who, in the late nineteenth century, referred to a large area of land as a "park". His understanding is that Central Park in New York was created to provide a sense of greenery and the impacts of a rural landscape (ibid.). In the words of Richard Murray:

> "Demand for large-scale urban parks emerged at the height of the First Industrial Revolution in the mid-1800s, when large-scale urban parks represented new ideas of accessible public spaces, often established on land previously owned by aristocracy, royalty or the army. They represented new ideas on how city life could be improved and how large green spaces could enhance urban citizens' physical and psychological well-being." (Murray 2022)

But times have evolved. The static, pastoral ideal represented by the park model from the nineteenth century is no longer relevant. According to journalist Arthur

Lubow, "The Anti-Olmsted" in *New York Times Magazine* in 2004 is the design model for expansive urban parks on post-industrial lands. In the same year, landscape architect Alan Tate criticized this type of park as being iconic, ageless, and timeless without process and change taking place in her book *Great City Parks* (for instance, Central Park in Figure 2 and Prospect Park) (Tate 2004). The book *Large Parks*, published in 2007, compiles a number of critical debates on this subject. It is not difficult to discover that the paradigm shift of parks is unavoidable in theory and practice when compared to a pastoral ideal for static park landscapes in an industrial society and a disorganized, chaotic urban reality for process-oriented park landscapes in a post-industrial society.

Why is the pastoral ideal questioned in the development of landscape theories? In a 1964 work entitled *The Machine in the Garden*, American historian and literary critic Leo Marx stated that "the pastoral ideal [ ... ] is located in a middle ground somewhere 'between' yet in a transcendent relation to the opposing forces of civilization and nature" (Marx 1964). This suggests a struggle between an idealized view of the natural world and industrialization, symbolized by technical development, which provides the pastoral ideal with a "counterforce" (ibid.). The park from the nineteenth century is considered a remedy for both the industrial metropolis and nature. Is there still a relationship between the city and the park, technology and nature being in conflict with one another? In the different cultural contexts of the twenty-first century, it is obvious that conflict is replaced by a more complex, dynamic, and integrated interaction.

**Figure 2.** Central Park in New York, Frederick Law Olmsted and Calbert Vaux, 1857. Source: Photo by ©Linfei Zhang 2016; used with permission.

More importantly, intentionally preserving a park that displays similarly picturesque beauty on a global scale might prevent different people from understanding the landscape. Cities, landscapes, nature, and cultural ideas are all changing. Landscape architects have always understood that their works and ideas develop through time. Peter Latz once said that "ideologies change faster than trees can grow" in reference to this (Latz 2016b). What kinds of aspects of contemporary large-scale urban parks should be theorized by experts in the fields of landscape architecture, urban ecology, and urbanism in order to improve the understanding of parks? This is a problem that all professionals must deal with.

As indicated above, professionals' increased understanding of the *dynamic landscape* has led to the opening up of a variety of landscape research avenues, which has enhanced the concepts of parks, their design paradigms, and the general perception of contemporary urban landscapes. The book explores the evolution of park landscapes from a variety of cultural perspectives as well as the driving forces behind contemporary urbanism and landscape architecture.

In reality, there have been a lot of theoretical investigations and critical analyses of expansive urban parks. Among them is a large international conference called "The Large Parks: New Perspectives Conference," which was held in GSD as early as 2003 and whose critical analysis largely identified the paradigm shift of large-scale urban parks. It was intended to overcome a long-standing "Either-Or" dilemma or traditional duality of thought in the field of landscape architecture, which will be covered in Chapter 4.

As a result, in the 1990s in North America, the field of landscape architecture produced a revolutionary park model known as *large parks*. The concept was developed by North American professionals, including James Corner, Julia Czerniak, George Hargreaves, and Nina-Marie Lister, via the lens of size. The idea of *large parks* in North America is essentially an organic design paradigm that reflects post-industrial and "extensive landscapes" (Corner 2007) and responds to the "critical" "professional" reformulation.

As a result of Scottish landscape architect Ian McHarg's *Design with Nature* from 1969 and J. B. Jackson's understanding *of dynamic landscape* from 1984, we refer to these kinds of *large parks* as "organic parks" most frequently in this book. The classic and ideal static nineteenth-century conventional park model created by Frederick Law Olmsted is overstepped in terms of function and space. The organic park model is a practical application of the *landscape urbanism* program, which is a design concept for biological processes in urban landscapes. Based on changes in the broader socio-economic structure and the present understanding of nature and ecology, this North American template demonstrates how reclaimed industrial areas are changing.

Germany's decoding, understanding, and representation of a physical site design philosophy led to structuralistic parks with material structures (Rosenberg 2007). Researchers found that the stereotypical reproduction of outmoded nature and landscape images was not the way ahead as indicated by using the critical approach (Weilacher 2008). In order to criticize "the conventional approach of wanting to preserve the industrial relics merely as alienated, incomprehensible monuments, as aesthetically attractive curiosities, without attempting to tie them into the complex landscape context," German structuralistic parks are established on post-industrial sites. Importantly, the structure is seen as an analytical method, designers' syntax, and a spatial carrier for the growth of park landscapes. Furthermore, the syntactic structure contains existing material and information on site and designers' own perception and cultural interpretations, which can absorb spontaneous interventions without changing the essence of the scheme. They can even be a welcome variance. In essence, the structure of the park is the product of perceiving the complex interactions between humans and the natural world. In the face of an increasingly problematic environment of derelict sites, the structure can also serve as a medium and framework for shaping the material world and our thoughts.

The discussion of contemporary large-scale urban park models from North America and Germany showed how they considerably constrained the park model from the nineteenth century. The conceptual approaches of the two new models are seen as being distinct from one another and reliant on the material "structure" or idealistic infrastructure that they are considered upon. Regarding "structure" in space or "matrix" in landscape ecology, the concept of large urban parks is divided into three categories: North American model of organic parks, German model of structuralistic parks, and Chinese model of shan-shui parks, which are each interpreted differently in Chapter 4.

Through a critical analysis of typical examples of post-industrial landscapes in these three regions, this study explores and categories various park paradigms with essential urban park cases based on an extensive survey, as shown in Table 1. Some of the significantly influential park paradigm cases with large sized examples will be explored in this book. Even though there are not so many park projects, they signify new-emerging landscape thoughts in the field of landscape architecture and landscape ecology. Specifically, Parc de la Villette in France appears in the category of organic parks, because it is considered as the beginning of large-scale conversion of urban derelict land into a new park. It reflects complexity, creativity, and unpredictability, which inspired the creation of North American organic parks. Another special case is Gas Works Park, which was built in the early years of post-industrial landscape transformation. As one of the first examples of bio-remediation, it is not truly an organic park, but it may set the stage for the development of this park model.

For the structuralistic park paradigm, several large parks in this book are concentrated on Latz + Partner's projects, because of their success in shaping the post-industrial landscape in Germany and other countries (such as Dora Park, Italy). Peter Latz's design philosophy, structural syntax, and attitudes towards industrial nature and culture are the best in the transformation of large-scale urban parks. The structural analytical approach is thus considered the core of landscape planning and design for many German designers to deal with abandoned sites and green open spaces (such as Park am Gleisdreieck in Berlin and Olympic Park in Munich). The approach even exerts a profound influence on the Chinese landscape practice from aspects of the re-organization of regional space, hierarchical treatment of historic elements and structures, and means of minimal intervention, such as Beijing-Zhangjiakou Railway Heritage Park, Yangpu Waterfront Reconnection Projects. In fact, this book expects to provide professionals from a wider range of countries with a deep and detailed understanding of the nature and characteristics of the structuralistic approach, and it is by no means limited to merely describing Latz + Partner's post-industrial landscape projects.

**Table 1.** Three large-scale urban park paradigms with essential cases in the study areas.

| Study Area | Park Paradigm | Park-Paradigm Case | Post-Landscape Case |
|---|---|---|---|
| North America | Organic park | Parc de la Villette, Gas Works Park, Pyxbee Park Freshkills Park, Downsview Park | High Line, Philadelphia Rail Park, Domino Park, SteelStacks Arts & Cultural Campus |
| Germany | Structuralistic park | Landscape Park Duisburg-Nord, Bürgpark Hafeninsel, Dora Park, Park am Gleisdreieck, Olympic Park, Riemer Park | IBA Emscher Park, IBA Fürst-Pückler-Land, Heilbronn Ziegeleipark, Tempelhofer Feld Park, Bernepark |
| China | Shan-shui park | Shougang Industrial Heritage Park, Beijing Forestry Park, Shanghai Houtan Park | Beijing-Zhangjiakou Railway Heritage Park, Shanghai Minsheng Wharf and Yangpu Waterfront Reconnection Projects, Hangzhou Xiaohe Park |

Source: Author's compilation based on data from Weilacher 2008; Fabris and Li 2020; Li 2023.

In addition, according to the list, more large-scale urban parks on industrial derelict sites could be classified under the park paradigms in the future, but not limited to these three kinds. Actually, in the search for park paradigm cases, other post-industrial landscape projects of different scales and types are also clearly seen. On the one hand, they demonstrate the increasingly great need for the transformation of abandoned sites in urban areas, and on the other hand, the "park" becomes the core concept meaning open space for diverse social uses in a broader sense.

### 1.3. Large-Scale Urban Parks for Post-Industrial Site Conversion

Large-scale urban parks are increasingly being created on post-industrial sites, and according to landscape designer Julia Czerniak, whose work focuses on the physical and cultural potentials of urban landscapes, their limits, often political and economic as much as geographic, are imposed, not chosen (Czerniak and Hargreaves 2007). These site "limits," in the eyes of the landscape architect, give contemporary large-scale urban parks a special quality. In order to draw attention to the transformation of land-use types from industrial to post-industrial societies or landscapes as a tool for the reclamation of former industrial sites, the majority of these distinctive parks in Germany are referred to as *Postindustrielle Landschaftsparks* (*Post-industrial Landscape Parks*) or *Landschaftsparks* (*Landscape Parks*).

However, both German conceptions of large urban parks fall short, especially when compared to the re-comprehended urban landscapes in this study. Instead of using the term *"post-industrial landscape parks"*, large-scale urban parks are used to close off the study dimension of German structuralistic parks from a size viewpoint. Additionally, the word "large" denotes a range of meanings. The size primarily symbolizes the inclusivity and potential of the post-industrial "thinking" park models, with a focus on managing complicated and hazardous sites. An increasing number of landscape architects are concerned about how parks on these sites are typically made, created, and cultivated-designed from "more open-ended processes and formations" (Corner 2007). For example, during the regeneration of Beijing-Zhangjiakou abandoned railway, this open-ended process means the vegetation succession and self-growth of a new linear urban park in order to maintain and develop the original wilderness character of the site, as shown in Figure 3. The park is finally constructed with the design idea of integration, connection, sustainability, and diversity for developing urban nature and public space. The post-industrial landscape in this park structure thus means a specific spatial scene containing historical information and diverse landscape elements.

**Figure 3.** Beijing-Zhangjiakou Railway Heritage Park, Xiangrong Wang, 2019. Source: Photo by author (right) and ©Danzi Wu 2019 (left); used with permission.

In addition to size, the word "large" connotes the "ambition" of North American researchers to create a conceptual framework that connects urban form, dynamic environmental processes, and daily living (Czerniak and Hargreaves 2007). The term "large" for organic parks refers to the complexity and resilience of the ecological system as if it were a creature. It is considered by German academics as a "large thinking" park model for the entire area. As a result, these parks were created for specific, single sites, such as IBA Emscher Park (1989–1999) in the broader Ruhr region. German structuralistic parks without boundaries can be seen in this light as a strategy and process for the slow and methodical rehabilitation of urban regions.

Large-scale urban parks for post-industrial site conversion are considered in this book as a key concept in the field of landscape architecture. It proves that society, nature, time, and space are always interconnected in the development of a park's landscape. Large-scale urban parks connect the history, present, and future of cities, observe the development of society and the expansion of urban space, and incorporate a variety of interactions between humans, nature, and cities. Large-scale urban parks serve as both a nature and cultural reserve and as visions of urban green, quality of life, cultural and historical preservation, as well as ecosystem services and social cohesion (Murray 2022). The term "large-scale urban parks on post-industrial sites" in this study encompasses several crucial elements, including:

- As a kind of green idea, large-scale urban parks reflect the natural consciousness and the comprehension of cultural landscape in different regions;
- As a social space of 'urban nature', they ensure high-quality, healthy and shared living spaces for people in their diverse urban lives;
- As 'green infrastructure', they address climate change and ecological problems;
- As a 'strategic concept' for contemporary urban landscape planning, they may influence regional landscape transformation and revitalization and represent the power or potential of the landscape, among which some keywords can be discovered, such as resilience, openness, uncertainty, inclusiveness, process orientation, etc.

Studies on large-scale urban parks as a crucial form of the post-industrial landscape are conducted in urban areas in response to the potential and challenges of abandoned industrial sites. The expansion of expansive urban parks as a fundamental concept promotes urban renewal and the redevelopment of polluted and abandoned industrial sites, both in principle and in practice.

How can we understand the cultural significance of large-scale urban parks in developed and developing nations from a global viewpoint within the context of contemporary urban landscapes? In order to perform an in-depth and cross-cultural examination, this book will focus on different design paradigms for large urban parks in North America, Germany, and China. Its goal is to explore the pluralistic understanding of contemporary landscapes at the urban scale and diversified post-industrial landscapes in the form of large urban parks.

# 2. Large-Scale Urban Parks in Critical Thinking

"Flows, connections, and assemblages replace the concept of borders; and there is a growing interest in the type of connection, its components, its spatial expression, and the social and cultural processes related to it. These shifts have a far-reaching influence not only on our culturally shaped perceptions of cities, nature, bodies, countryside and landscape, and their corresponding images but also on design approaches." (Giseke 2018)

How should large-scale urban parks be treated, when the contemporary urban conditions are in constant flux? What methodologies are applicable to guide the rethinking of large-scale urban parks in landscape architecture academia? If the nineteenth-century urban park concept proposed by Frederick Law Olmsted is taken as an ideal landscape model for the urban environment of that time, what would a specific, new large-scale urban park model in the twenty-first century be like?

## 2.1. Transition in Cities

From a dynamic perspective, cities are evolving and becoming interconnected with the flows and exchange of materials and energy, as reflected in the ideas of cities, such as the original recognition given by German philosopher Karl Marx in the nineteenth century to the dynamic internal relationships between humans and nature, the "metabolism of cities" put forward by American engineer Abel Wolman (1965) based on a biological metaphor of living organism, and the "dynamic system of manmade spaces" proposed by J. B. Jackson in 1984.

The flows of materials, information, energy, and "society-nature interactions at different spatiotemporal scales" (Haberl et al. 2019) in the transition of cities demonstrate the processes of changes in spaces, society, and nature and lead to progress in people's knowledge, experience, and planning and design approaches to urban landscapes. In other words, complex urban spaces with the characteristic of transience, mobility, cirularity and exchangeability have changed the comprehension of contemporary cities grasped by urban planners and landscape architects. As argued by the German scholar Alain Thierstein, dynamic cities not only create efficient, effective interaction but also an agreeable atmosphere and pleasant surroundings for the users, which requires a wider and more advanced interpretation of cities (Thierstein 2018). Furthermore, an increasing number of scholars consider that the integration of landscapes, settlements, and infrastructures produces the complex contemporary urban landscape, which is composed of the interaction of spaces of mobility and spaces of place, as shown in Figure 4.

**Figure 4.** The contemporary dynamic urban landscape, Beijing. Source: Photo by ©Ziyue Wang 2020; used with permission.

The essence of dynamic cities has prompted extensive exploration in space-related research areas. As German urban designer Sophie Wolfrum stated, new spatial forms of cities together with new urban landscapes are emerging (Wolfrum and Nerdinger 2008). The reason for the production of new spatial forms lies in physical changes in complex urban realities, including the dissolution of urban structure in the spatial dimension and the growth of urbanity in the social dimension.

Hereby, the landscape readjustments at the urban level in the field of landscape architecture should be treated from the perspective of revised cities. The thought of connecting changing cities with conceptual urban landscapes is indicated by German landscape architect Sören Schöbel as "views from the outside of landscape by itself" (Schöbel 2007).

### 2.1.1. Changing Cities

Under the profound influence of irresistible urbanization and globalization, "the world's cities are changing" (Wolfrum and Nerdinger 2008). Generally, contemporary cities around the world are subject to such changes: growing and shrinking, and flourishing and declining (Schäfer 2005). Those in developed countries, such as countries in North America and Germany, are experiencing urban shrinkage and perforations as a result of "the decline in population and the closure of industrial installations" (Dettmar and Weilacher 2003). In contrast, those in developing countries, such as China, are undergoing rapid and massive urban expansion.

With the shrinking of cities in developed regions, the term *post-industrial society* was first put forward by French sociologist Alain Touraine in 1969, which is closely associated with some similar theoretical concepts of sociology, such as *Post-Fordism, Knowledge Economy,* and *Network Society.* The notion of *post-industrial society* was popularized by American sociologist Daniel Bell in his 1974 work *The Coming of Post-industrial Society.* It reflects that the huge transition of society brings about the issue of post-industrial site transformation with the process of deindustrialization, which is regarded as a common challenge in today's landscape architecture.

Consequently, the increasingly prominent issue of managing sites' disorder and complexity elicited critical requirements for large-scale parks in metropolitan areas instead of the antiquated pastoral park paradigm in former industrial cities. Canadian ecological designer and planner Nina-Marie Lister, whose research focuses on the intersection between landscape infrastructure and ecological processes in metropolitan areas, indicated that cities are "revitalizing their post-industrial areas, often through the creation of large urban or exurban parks" (Lister 2007).

In other words, the huge transition around the world from industrial to service economies created a vast inventory of large abandoned sites (Corner 2007), contributing to the stimulation of the development of large-scale parks on such "disturbed sites", including "quarries, water-treatment facilities, power-generation plants, factories, steel mills, landfills, military bases and airports" (Meyer 2007). For instance, the old Oberwiesenfeld military base intensely utilized up until 1939 was repurposed into Olympic Park in Munich with the theme of open-space-guiding urban development. By making full use of the long-standing rubble on the site, the project reclaimed what was previously a mountain of waste into an undulating, spatially varied area of green hills. The design team that planned this large-scale urban park considered and then creatively expressed the structural relationship between the Alps and Munich city, and a full compact system of blue–green space was thus formed for further sustainable spatial development, as illustrated in Figure 5. In conclusion, changing cities as an urban phenomenon in developed regions prompted the emergence of post-industrial areas, making large-scale urban parks an instrument to activate derelict sites.

**Figure 5.** Munich Olympic Park, Günther Behnisch and Frei Otto, 1972. Source: Photo by author.

However, most developing cities in China retain rapid urban growth and expansion. Beijing, as the first city to engage in country park planning and design in suburban areas, is taken naturally as an example of a changing city in China. As per the Master Plan of Beijing 2002–2020, the overall urban built area increased with the expansion of the ring-shaped infrastructure system from 1975 to 2002 (Stokman et al. 2008). Such a plan suggests that rapid population growth and an increase in housing and economic activities have expanded the inner city's built-up area from 84 km$^2$ in 1949 to more than 700 km$^2$ within the recent decade since the foundation of the People's Republic (Li et al. 2005b). One of the impacts of economic reforms in 1978 was unprecedented urban expansion. The distribution of settlements and infrastructures followed a dominant ring-road system; the second and the third ring roads were established separately from the 1980s to the 1990s. The apparent urban growth is another embodiment of changing cities worldwide.

Regardless of different urban situations, Beijing city is now in the process of moving towards becoming a *post-industrial society*. In this move, the issue of remnant and derelict industrial land is considered as the relocation of industrial enterprises follows industrial restructuring. These enterprises include the Shougang Group, one of China's largest steel companies in Beijing city, and Beijing Coking Plant. Other considerations consist of the demand for improving urban eco-environmental conditions and the integration of fragmental spaces in the urban–rural fringe. With the establishment of Shougang Industrial Heritage Park, the landscape has become a structural tool for restructuring the whole region, integrating resources of green open spaces and rehabilitating industrial wastelands, as exhibited in Figure 6. In the structural analysis of space and architecture, the black color stands for the large-scale Shougang Park and its surrounding green open spaces in an urban system.

**Figure 6.** Shougang Industrial Heritage Park in Beijing, Tsinghua Urban Planning and Design Institute, 2016. Source: Photo by author.

Furthermore, the relocation of the Shougang Group exerts an influence on the analysis and conception of Chinese urban landscape and large-scale urban parks. While transforming gradually into a *post-industrial society*, Beijing city will meet challenges similar to those met by the two developed regions. More Chinese large-scale parks with a kind of morphological feature of shan-shui will emerge in these abandoned industrial sites to cope with these challenges.

In the above consideration, the nature of dynamic cities in countries in North America, Germany, and China results in different levels of urban development. All the same, there are still regular changes in the morphologies of cities, according to the sophisticated experience of developed cities. From the viewpoint of theoretical analyses, they are concluded through several essential urban models, suggesting that "the transformation of ordered cities into urbanized regions" as a global urban phenomenon is occurring equally in both developed and developing countries (Sieverts 2008). These inductive urban models "represent ideals of what some people think a city ought to be" while undoubtedly reflecting a relatively direct-viewing analytical method (Shane 2011).

In this sense, British architect Cedric Price put forward "three city morphologies in terms of breakfast dishes" in 1982. Graphically described as "boiled" (ancient), "fried" (seventeenth–nineteenth century) and "scrambled" (modern), his "three eggs diagram" demonstrates a shift from the traditional and dense city fixed in concentric rings of development within its walls to a postmodern city, where everything is distributed uniformly in small units to constitute a continuous network (Shane 2006). For example, the traditional city of Entrevaux in southeastern France remains protected by defensive walls, and the modern city of Munich is surrounded by infrastructure networks and landscape, as shown in Figure 7. This diagram assumes general patterns to express the diffusion of a city in the course of space and time. In 2001, the young planners' group of ISOCARP renewed these three categories of urban models of "walled city", "center + agglomeration" and "periurbanisation", ascribed to the increasingly complex and fragmented urban spatial environment (Shane 2011), as can be seen from Figure 8 and Table 2.

**Figure 7.** Examples of the walled city of Entrevaux in France and the modern city of Munich in Germany. Source: Photo by author.

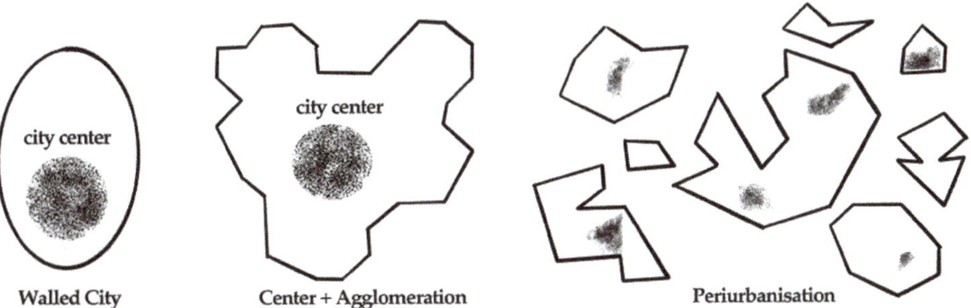


**Figure 8.** Illustration of three categories of urban models according to ISOCARP. Source: Figure by author.

**Table 2.** Three renewed urban models, their main features, and related specialists and tasks.

| Walled City | Center + Agglomeration | Periunibanisation |
|---|---|---|
| 1. Agricultural production; 2. Muscular movement; 3. Feudal government | 1. Industrial production; 2. Mechanical movement; 3. Democratic government | 1. Informational production; 2. Light-speed communication |
| 1. Architect; 2. Local scale; 3. Architecture, boulevards projects | 1. Physical planner; 2. Regional/national scale; 3. Land use, infrastructure plans | 1. Spatial development manager … ? 2. Knowledge + information infrastructure 3. Strategies/action.? |

Source: Table by author.

The general construction of urban models helps to visualize shifting morphologies of cities, while these models are not enough for fully theoretical analyses of a series of deep changes related to urban space and society in developed regions. Hence, more analyses should be conducted.

### 2.1.2. Dissolved Urban Structure and Growth of Urbanity

From the perspective of urban space, evident distinctions between city and landscape as well as urban and rural areas among developed countries had largely blurred at the end of the twentieth century (Bruegmann 2008). The situation of a dissolved urban structure has been analyzed by numerous professionals. As early as the 1900s, British historian and novelist Herbert George Wells predicted "the probable diffusion of cities" in his 1902 book *Anticipations*. He envisaged that "these coming cities will not be, in the old sense, cities at all; they will present a new and entirely different phase of human distribution" (Wells 1902).

However, since the beginning of the last century, European professionals have verified the reality of the prediction of decentralized urban regions (Wolfrum and Nerdinger 2008) with the periphery of the city in urban development and even suburbanization and the massive expansion of infrastructure for mobility and production (Höfer and Trepl 2010). As German sociologist Detlev Ipsen, who focuses on urban and regional sociology, expressed:

> "Sophisticated goods, leisure facilities or workplaces are no longer predominantly concentrated in the central city, but in the urban region [ … ], one consequence is that the classical 'center-periphery' commuting pattern is displaced by more diverse networks." (Ipsen and Weichler 2005)

The concept of the *urban region* emerged, which signifies that perforating cities have perforating urban landscapes, as shown in Figure 9. In the view of Thomas Sieverts, this concept in European academic circles is indeed a new urban form called *Zwischenstadt*, which addresses the decentralization of the compact traditional European city and examines the new form of urbanity that has spread across the world. *Zwischenstadt* is "neither city nor landscape" (Sieverts 2003). He suggested that "the city is integrated with the landscape, and the old contrast between town and country has already substantially dissolved in favour of a city-landscape continuum" (ibid.). The "city-landscape continuum" not only demonstrates that the landscape no longer lies outside the city, while the city no longer lies in the landscape (Dettmar and Weilacher 2003), but also implies that the city and landscape, the urban and the rural, increasingly demand being commonly considered.

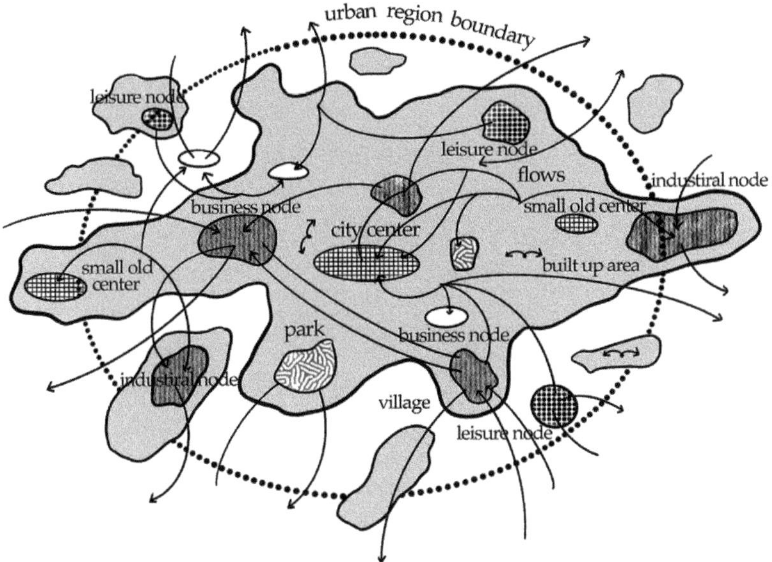


**Figure 9.** Illustration of urban region in the view of Ipsen. Source: Figure by author.

Moreover, demonstrating the dissolved urban structure and growth of urbanity also present in the form of *porosity*, the perforating urban landscapes in urban regions originated from the term *Porous City*, as proposed in 1925 by German philosopher and critical theorist Walter Benjamin and Asja Lacis. The *Porous City* is regarded as one of the contemporary urban models showing permeability in space and society. In Europe, it sparked wide-ranging discourse in such sectors as architecture, urbanism and landscape architecture, to inspire "a progressive urban agenda" (Wolfrum et al. 2018). In Sophie Wolfrum's words:

> "The porosity refers to the ambiguous zones, in between spaces and thresholds that permeate urban environment. Such spaces merge into each other, providing the backdrop for the unforeseen and improvised, and blur the boundary between physical and social space." (ibid.)

On the one hand, there is much significance attached to the morphological characteristic of porosity with respect to landscape architecture. As an essential property of spatial boundaries in both nature and landscape, porosity ensures the connection of separate landscape units to each other and to the environment (Weilacher 2018). On the other hand, the natural and manmade infiltrations in porous cities can breathe life into the city based on the "social production of space" put forward by

Henri Lefèbvre (Bauer et al. 2018). The physical and social spaces in urban regions are the routine elements of urban landscape analysis. For instance, in the contemporary model of the *Porous City*, landscape and urbanity expand inside and outside the city. As shown in Figure 10, the urban texture of Helsinki reflects the characteristic of porosity where blue-green spaces are vitally important.

**Figure 10.** Aerial view of Helsinki in Finland reflecting the porosity of contemporary European cities. Source: Photo by author.

Furthermore, the description of the *Porous City* prompts more and more scholars focusing on the city scale to explore the correlation between the urban model, urban landscapes, and post-industrial sites. For instance, Italian architect and urbanist Bernardo Secchi made reference to the original interpretation proposed by Walter Benjamin and Asja Lacis to argue that it is necessary to rethink and redesign the urban landscape as "a new urban form is arising" (Secchi 2007). With the European city Antwerp used as a case study, he discovered that the "abandoned industrial sites and buildings within the urban fabric have turned Antwerp into a *Porous City*. The porosity offers the opportunity to create a new constructed landscape within the urban region" (ibid.).

In summary, as demonstrated by the urban models of the *Zwischenstadt* and *Porous City* in Europe, contemporary cities have been transformed to a large extent, which prompts us to interpret them from such perspectives as morphology, landscape, functions, society, and the way of governance. Among them, landscape determines the future development of the city (Weilacher 2018). By discussing the model of cities, the key viewpoint is supported that the change in urban structures is definitely correlated with that of conceptual urban landscapes.

However, contemporary cities would be largely developed into homogeneous entities if the analysis was merely laid on dissolved urban structure with the consideration of the dynamic nature of urban space. With the impact of globalization or *Fordism*, the mobility of urban spaces occurs inescapably, creating a new urban form to a certain extent. Such form is characterized by dissolved urban structures with disappeared polarities, embodying a new mix of people and land in a specific society. The new urban form transforms into a different city when it is endowed with social meanings, including a distinctive urban lifestyle and diverse social relations as well as cultural value. This is the significance of growing urbanity.

In this regard, the growth of urbanity is another aspect of defining contemporary cities. Interpreted from this view, the dissolution of urban structure signifies that old rural lifestyles disappeared as "most people's spheres of life have long overstepped the boundaries of the local community and have extended to the whole urban region" (Sieverts 2008). The comprehension of urbanity is essentially influenced by Chicago School's urban sociologist Louis Wirth's idea of "urbanism as a way of life" (Wirth 1938). His idea reveals that "the city is wherever an urban lifestyle is" (Dettmar and Weilacher 2003).

Due to the urbanity in the critique of a functional separation of spheres of life, the academic discourse took place within urban studies, such as Jane Jacobs' *The Death and Life of Great American Cities* in 1961, which involves social space for complex and diverse uses; American architectural historian and theoretician Colin Rowe and architect and urbanist Fred Koetter's *Collage City* in the early 1970s, which emphasizes "mix use, superposition, ambiguity, urban diversity, potential spaces and collage"; and German sociologist Hans Paul Bahrdt's *The Modern Metropolis* (German: *Die Moderne Großstadt*) in 1961, which is based on sociological thoughts on urban development and aspects of urban life (Wolfrum 2018).

In conclusion, the revised understanding of contemporary cities is outlined clearly based on the aforesaid arguments on physical change in the urban environment. The semantic shift in cities makes analyses and concepts of contemporary urban landscapes enter into a renewed stage. Contents of this stage could be explained critically as two different theoretical *schools of thought* about contemporary urban landscapes.

### 2.2. Critical Rationalism Method

The physical changes in the urban environment have triggered the reflection on the landscape at the urban level in North America, Germany, and China. Hereby, critical rationalism is considered a scientific research approach to studying contemporary urban landscapes and large-scale urban parks. In the 1957 book *The Poverty of Historicism*, Karl Popper used the term critical rationalism to indicate a modest and self-critical rationalism. The term was derived from rationalism because he agreed with German philosopher Immanuel Kant's philosophical system of rationalism during the eighteenth century, which stated that human rationality creates "laws of nature" (Rohlf 2008).

However, Karl Popper questioned the widespread correctness of *rationalism* and thus moved his critical rationalism toward "falsifiability" (Popper 1976). In *Unended Quest*, he posed the following question to suggest "the logic of scientific discovery" (Popper 1959) and the "falsifiability":

"My main idea in 1919 was this. If somebody proposed a scientific theory he should answer, as Einstein did, the question: 'Under what conditions would I admit that my theory is untenable?' In other words, what conceivable facts would I accept as refutations or falsifications, of my theory?" (Popper 1976)

The questions worth pondering illustrate that a universal theory is never eternal or enduring. Instead, it is open to continuous questioning. Additionally, even a theory is scientific only when it has a probability of falsification, which makes significant sense to Karl Popper. For him, falsifiability is "a criterion of demarcation" used to distinguish between "science and pseudo-science" (ibid.). In conclusion, theories move forward through ongoing falsification, negation, and criticism.

Moreover, in respect of urban spatial and social structures, he directed criticism at the idea of utopian social planning on a large scale as an illusion luring us into a swamp (Popper 1957). The "falsifiability" in reality can be used to shatter the utopian model and its grandiose plans. Karl Popper's critical rationalism exerts a profound influence on sociological and urban planning concepts (Schöbel 2018), from *Collage City*, Karl Ganser, Walter Siebel, and Thomas Sieverts' *Perspective Incrementalism* (German: *Perspektivischer Inkrementalismus*), equivalent to muddling through; Willem van Bodegraven's *Structuralism* in architecture in 1981; André Corboz's *Territory as Palimpsest* in 2001; to Ulrich Beck's *Reflective Second Modernism* in 1993.

In the field of landscape architecture, transferring the philosophic approach of critical rationalism to the research and planning approaches generally guides us throughout our reflection on contemporary large-scale urban parks and urban landscapes. Karl Popper's critical rationalism implies that the classic nineteenth-century park model demands to be theoretically contradicted with the advancement of society, despite its existence in reality. The 'untenable' park design paradigm gives rise to the following research question:

How are contemporary large-scale urban parks regarded within changing cultural conditions, in terms of urban spatial structure, society, and ecology?

## 2.2.1. Two Critical Thinking Approaches

To answer the aforementioned question, the critical rationalism approach is divided into two meanings to consider North American and German urban landscapes and large-scale urban parks. In other words, these analyses and conceptions are specifically defined as critical rationalism approaches, which are primarily manifested in *critical thinking* proposed by James Corner and *critical structuralism* interpreted by Peter Latz. Both views embody a critical, professional perspective in analyzing the current landscape architecture.

In comparison with these two existing approaches, *critical thinking* in the Chinese urban landscape and large-scale urban parks remains in the exploratory stage. This is attributable to the lack of contemporary urban landscape theories in Chinese landscape architecture academia. It is thus necessary to establish *critical thinking* on the landscape in a specific cultural context. Therefore, this book is intended to shed light on contemporary landscapes in China at a personal level, which is based on explaining the critical landscape analyses and formations in developed regions.

Above all, James Corner's *critical thinking* is interpreted as the critical rationalism approach to North American urban landscapes and organic parks, because he took the lead in bringing a critical perspective to the discipline of landscape architecture in the early twentieth century, influenced by J. B. Jackson's landscape concept analysis in the 1980s. He proposed ideas in his 1991 essay Critical Thinking and Landscape Architecture:

> "Critical thinking begins with skepticism, particularly with regard to authority, rules, and conventions that have long gone unquestioned. [...] Critical thinking also involves reflection, a considered and thoughtful analysis of the issues and values involved. This is followed by speculative contemplation, a formulation of alternatives and possibilities—necessarily fluid and unconstrained. Finally, critical thinking culminates in action: decisions are made, and work is done." (Corner 1991)

James Corner's *critical thinking* aims toward creative action. In this viewpoint, today's *critical thinking* is supposed to be more about "the creative processes of making and action than it is about theories of theories" (ibid.). The critiques of theories per se clearly do not embody his understanding of the critical rationalism approach to the urban landscape. Moreover, the creative processes emphasized by James Corner actually coincide with his idea about North American landscapes in *cultural imagination* (Corner 1999). *Critical thinking* reflects creativity in action, in the aspect of culturally re-interpreting landscapes.

For James Corner, the creative processes are represented by his unique operational method called 'plotting' for the practical conception of complex, dynamic sites, as specified in the Chapter 4. In his mind, landscape architects are "plot-makers", who make plans, stake out and delineate territory, and unfold the passages of time, through activities of "digging, surveying, mapping, planning, founding, shaping and drawing" (Corner 2021).

For example, High Line was designed in this way by Diller Scofidio + Renfro and James Corner's Field Operations for the transformation of a nine-meter-high disused viaduct into a simple, quiet, and wild promenade. Plots are bound to uncertain life. According to James Corner's critical thinking, "plots have life; plots shape life; and plots instigate life" (ibid.). In the design idea of keeping simple, wild, quiet, and slow, the High Line thus unfolds the interplay of plots, life, and nature, as shown in Figure 11. In conclusion, the cultural embedding of creativity is certainly reflected in the understanding of the North American critical approach.

**Figure 11.** High Line in New York, Diller Scofidio + Renfro and Field Operations, 2009. Source: Photo by ©Linfei Zhang 2016; used with permission.

Furthermore, the critical rationalism approach to German urban landscapes and structuralistic parks is manifested in different planning styles developed since the early 1980s. An example is the *Perspective Incrementalism*, which is used at the IBA Emscher Park and in the same surrounding of Peter Latz's method. This is referred to as *critical structuralism* in this book. The concept of *Perspective Incrementalism* was shaped by IBA Emscher Park leader and managing director Karl Ganser, German sociologist Walter Siebel, and Thomas Sieverts during the 1980s: "mit dem vorgestellten Adjektiv ist die Vielzahl der kleinen Schritte gemeint, die sich auf einen perspektivischen Weg machen" (Ganser et al. 1993). These German words describe a multitude of small individual measures that are, however, oriented toward an overall mission statement. The literature on *critical structuralism* in German landscape architecture is scarce, while Peter Latz's explanation meets its core. In an interview in 2016, he stated in German:

> "[ ... ] unserer Methode auch der kritische Rationalismus: Planung muss nicht nur verifizierbar, sondern vor allem falsifizierbar sein. Das muss einem als fester Bestandteil im Blut liegen. Das ist nicht einfach, denn wir befinden uns in einer Gesellschaft, einer Planung im Überfluss, und zwar einen Überfluss an Informationen." (Latz 2016a)

Peter Latz's statement implies that the critical rationalism approach guides landscape planning and design to be both verifiable and falsifiable, and should be considered an integral part. It is not easy for professionals to make falsifications and criticism, because we are in a society, planning in abundance, and indeed with an abundance of information. Nevertheless, the critical rationalism approach is expected to be grasped by them. Through this approach, German urban landscapes have been critically reconstructed and gently renewed since the 1980s, rejecting the radical modernist approach of rigid functional division. The German *landscape structuralism* movement emerging from this concept has affected the comprehension of the structuralistic parks.

The *critical structuralism* is a concept of structure pertaining to the characteristic urban landscape, deeply rooted in a unique cultural contextualization. The structure signifies complex, constructed, and layered landscape systems (Weilacher 2014). Accounting for the cultural contextualization, Peter Latz remarked that "landscape is basically history" that could not be "obliterated" but turned into "your partner" (Latz 2015). Hence, the approach emphasizes seizing "visible" and "invisible" "layers of information and elements" from the surroundings, keeping nearly everything for recycling, and then incorporating them into the structure (Latz 2008a). The reason is that "every element can become an element of the landscape" (Latz 2013a).

To sum up, compared with James Corner's *critical thinking* with creative processes in the *cultural imagination*, Peter Latz's *critical structuralism* with the structure embeddedness is treated as the method for criticizing generic urban landscapes without cultural contextualization. The goal of keeping everything for reinvention is realized through this approach. Peter Latz stated that the method is "between preservation and change" (Latz 2005). Through critical thinking, the cross-cultural comparison and communication of large-scale urban parks are conducted.

## 2.2.2. Three Park Models in Critical Thinking

For comparison, discovering global challenges and common tasks that the overall profession of landscape architecture must handle is a task against the background of ubiquitous urbanization and globalization. With the rise of suburban areas and the demand to cope with the ecological and environmental conditions after deindustrialization, the prominent global challenge in our discussion is related to the issue of site transformation, as well as sustainable urban regeneration and development vis-à-vis sociocultural and ecological considerations. Therefore, the common task is to realize the conversion of sites, especially contaminated industrial sites, through large-scale urban park planning and design in the conduct of spatial and temporal development.

However, distinguishing among park conceptions, conceptual approaches, and strategies is manifested in various responses under the common challenge. This is the primary aim of cross-cultural comparison and communication. Based on regional cultural diversity, these distinctive responses stimulate the analysis and development of contemporary urban landscapes and large-scale urban parks in their respective cultural contexts. The comparison can also fulfill the possibility of constant, extensive communication, and discussion in the field of landscape architecture.

Three design paradigms of contemporary large-scale urban parks are used in this research, and two of these come from developed regions: the North American organic model, which has been applied since the early 2000s, and the German structuralistic parks model, which has been explored since the late 1980s. Both models are illustrated based on the renewed understanding of contemporary urban landscapes in their critical thinking. Specifically, the theoretical formulations are divided into two branches: the 1990s North American *landscape urbanism* and the 1980s German *landscape structuralism*. Referring to these theoretical explorations and experiences, the Chinese shan-shui park model, including country parks as a typical peri-urban park implemented since the late 1970s and other landscape parks on post-industrial sites still inheriting the design concept of shan-shui culture in recent years, could be reflected within the conception of Chinese urban landscapes.

### Anti-Nineteenth-Century Park Model

Since the conventional understanding of nineteenth-century parks is considered 'untenable' in the postindustrial age through the critical rationalism approach, the evolving cognition of contemporary large-scale urban parks is expected to be established.

The classic park is identified as a generic, pastoral model, "borrowed from popular eighteenth-century landscape painting" (Weilacher 2008) and influenced by the traditional conception of "picturing landscape" (Waldheim 1999). On this basis, the American artist Robert Smithson said, "the 'pastoral,' it seems, is outmoded" (Smithson 1996). He demonstrated that the pastoral park model, expressing its conflict relationship with industrialization and technology at a particular moment in history, is actually outmoded (Rosenberg 2007). In this situation, the park is a "counterweight to an urban and industrial society" (Eisel 1982; Höfer and Vicenzotti 2013). Central Park in New York represents the ideal nineteenth-century large-scale urban park model and a green oasis. It has long been Inant in landscape theory and practice, as shown in Figure 12. This park model has been adopted by landscape architects across the world. If it is acknowledged that every culture makes an impact on the Earth's surface or urban landscape, and its own character is displayed (Körner 2013), it is necessary to answer several crucial questions in the process of critical thinking:

- Does this ideal, undisturbed image of urban nature need to be present in every city today? And how will the local, cultural expression of the urban landscape fit into it?
- Can this nineteenth-century park model be theoretically compatible with the ideas of changing cities and various urban landscape conceptions, and match the site characteristics and atmosphere of derelict land in post-industrial cities?

**Figure 12.** Central Park as a long-standing, worldwide paradigm of traditional large-scale urban parks, New York. Source: Photo by ©Zhenkun Gan 2019; used with permission.

However, the process of deindustrialization evokes the rethinking of contemporary parks, particularly in former industrial spaces and the re-imagination of relationships among parks, nature, society, and technology. Urban–natural, social, and technological factors are incorporated into contemporary large-scale urban parks, instead of the counterweight reference. They are naturally linked with the research question on how to regard contemporary large-scale urban parks in changing cultural conditions.

Moreover, the research hypothesis involves two large-scale urban park models from the developed regions. These two models are constructed with the critical approach that embodies the rethinking and conceptualizing of parks in post-industrial cities. Two different methods, the North American organic and German structuralistic methods, address the complex and contaminated industrial sites to transform them.

On this basis, it is worth reconsidering Chinese shan-shui parks within the distinctive shan-shui culture. In China, the shan-shui theories and practices have been applied to find the appropriate locations for human occupancy by exploring the landscape and identifying the irregularity and asymmetry of mountains and waters. However, the question is, how can this knowledge inform the contemporary practice of landscape architecture in China? Relative to these two park design paradigms in the West that have already be adopted in many contemporary park projects, will the model of shan-shui parks remain applicable to the transformation of derelict sites in the current understanding of Chinese urban landscapes? What should Chinese landscape architects learn from the two developed park models in their own practice of shan-shui park construction?

North American Organic Parks

In North America, the "critiques of modernist architecture and planning" proposed by American cultural theorist and landscape designer Charles Jencks (1977) influenced the *landscape urbanism* program. Consequently, the term "landscape" is significant and "uniquely capable of describing the conditions for radically decentralized urbanization, especially in the context of complex natural environments" (Waldheim 2006). Many traditional examples of nineteenth-century urban landscape architecture integrate landscape with infrastructure—Olmsted's Central Park in New York and Back Bay Fens in Boston serve as canonical examples (ibid.). Unlike the traditional model, "large-scale infrastructural landscape" is currently applied to contemporary practices of *landscape urbanism* in North America, such as organic parks. In this context, criticisms on the "camouflaging of ecological systems within pastoral images of 'nature'", belonging to the classic nineteenth century park to "integrate landscape with infrastructure", necessitate the conception of complex, dynamic, and living ecosystems established in organic parks as "large-scale infrastructural landscape" (ibid.).

North American organic parks with the ecological approach are an emerging park model driven by dynamic processes. This is the main body of the research because of its positive rethinking of the urban landscape in responding to the global challenge, advanced ecological ideas with a contemporary interpretation of nature, and noticeable theoretical explorations through a range of park competitions. Ideas about organic parks are mostly advanced by some North American scholars, such as James Corner, Julia Czerniak and George Hargreaves, etc.

In essence, the concept of organic parks is created to realize "a truly ecological landscape architecture" associated with the ecological approach, as suggested by James Corner's approach of *critical thinking* (Corner 1997). For Corner, the "truly ecological landscape architecture might be less about the construction of finished and complete works, and more about the design of 'processes,' 'strategies,' 'agencies,' and 'scaffoldings'—catalytic frameworks that might enable a diversity of relationships to create, emerge, network, interconnect, and differentiate" (ibid.). The Freshkills Park planning and design in 2001 led by James Corner fully demonstrated his large park assumption of ecological landscape architecture in practical examples to illustrate this statement. Over time, the park shapes "an ecology of various systems and elements that set in motion a diverse network of interaction" (Corner 2006).

German Structuralistic Parks

Compared with North American organic parks, the German concept of large-scale urban parks has a *structuralistic approach*. The *critical reconstruction* movement was proposed in Germany, where the modernist approach to city planning, architecture, and landscape architecture was also criticized. Under this influence and with other analyses on fundamental urban landscapes, the conception of *landscape structuralism* has been considered in German landscape architecture since the 1980s. With the *critical structuralism* employed in landscape architecture, the German park design paradigm is chiefly studied in this context. Consequently, unique structuralistic parks emerged.

The search for site structure, or *syntax*, becomes an essentially analytical step for German large-scale park conception (Rosenberg 2007). Peter Latz elucidated his emphasis on the structure in park design while critiquing the image:

> "It is not the images, but the abstractions, schemata of information layers or single systems that are required for understanding structure. The images of perfect examples that aim at the semantic level no longer show how it should be done." (Latz 2008c)

In other words, the structure demonstrates how the park should be analyzed and planned. With a recovered landscape as a key issue of urban regional development (Gailing 2005), the German structuralistic park is a strategy for keeping and retaining the site's industrial presence to the greatest extent possible. This concept reflects Peter Latz's viewpoint of "design by handling the existing" (Latz 1993). Many physical materials of sites related to cultural history and memory are analyzed and organized into multi-layered systems through the *structuralistic approach*. For instance, after the removal of dense vegetation from derelict sites in Landscape Park Duisburg-Nord, Latz + Partner discovered the undulations of the original railway embankments. They are described as "a gigantic piece of land art" (Latz 2016b) and as physical materials integrated into structural systems (Figure 13). These accepted materials are applicable to building the structure in line with the idea of landscape architects. On this basis, "new places" of large-scale urban parks are "invented at the fault lines between what was destroyed and what remained, between structures" (Beard 1996). They may boil down to his park design philosophy of decoding, understanding, and representing the physical site.

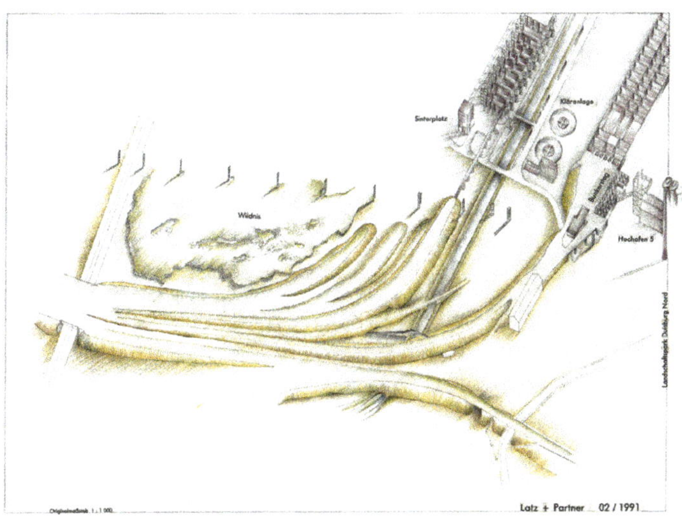

**Figure 13.** Landscape Park Duisburg-Nord, Latz + Partner, 1990. Source: Figure courtesy of ©Latz + Partner; used with permission.

## Chinese Shan-Shui Parks

The shan-shui parks proposed in this book as a significant method of park design for a shan-shui spatial structure can be integrated into the organization of traditional Chinese cities, parks, and gardens in line with shan-shui culture. Shan-shui parks are closely related to the ideal of the *shan-shui City* (the city of mountains and waters), a mega-structure for traditional urban construction, such as ancient Beijing, constructed by using a shan-shui structure at the urban level (Figure 14).

**Figure 14.** The shan-shui structure of ancient Beijing at the urban level. Source: Figure courtesy of ©Yun Qian; used with permission.

As the innate cultural essence in the profession of Chinese landscape architecture, shan-shui is closely connected to the traditional culture of emphasizing the unity of man and nature on a spiritual plane. This is subject to the profound influence from Chinese "Taoism" or "Daoism" ("道教" in Chinese), the literal meaning of which during the ancient Chunqiu period is "Teachings of The Way" (Rahmann and Walliss 2020). Chinese intellectuals have spent centuries contemplating the philosophy of nature, time, and space, which has led to an in-depth understanding of the human and non-human environment. Chinese designers have gained mastery as to the principle of Taosim, that is, man follows earth, earth follows heaven, heaven follows the Tao, the Tao follows nature, as argued by ancient Chinese thinker and philosopher Lao Zi. For the significance of Taoist thinking, the value of Taosim and its underlying ambition to harmonize with the rhythms of nature are viewed as the foundations for the development of a sustainable approach to landscape architecture (Chen and Wu 2009).

The term "shan-shui city" embodies the integration of urban construction with the natural environment comprised of various physical geographic elements. Embracing topographic and hydrological morphologies, that is, mountains (shan) and waters (shui), these morphologies are reflected as natural, site-specific, and man-made art in the course of planning and design. The shan-shui city places emphasis on the shaping of a regional spatial structure and shan-shui relations in the tradition of Chinese urban spatial planning. They are summarized into the formation of the shan-shui structure, which is a holistic approach to urban analysis and planning for ancient Chinese urban planners. In the meantime, it demonstrates one of the most crucial traditional and ideal philosophies, that is, the harmony between man and nature. As shown in Figure 15, shan-shui city embodies an ideal relationship between human and nature, mostly derived from traditional Chinese painting.

**Figure 15.** The Chinese ideal of shan-shui city borrowed from the image of a traditional Chinese painting. Source: Figure courtesy of ©Xiang Li; used with permission.

A strong perception of *first nature* or wild landscape is what reflects the plain shan-shui philosophical view of the landscape. For the Taoists, beauty can be reflected in this kind of nature, in the mountains and in waters, the quintessence of which can be extended into the symbolic microcosms of poetry, paintings, and garden design (Weller and Hands 2020). Through the shan-shui structure, the image of nature is abstracted and reproduced in the planning and design of parks. From the perspective of structure, the Chinese landscape appears to be regarded as both object and ideal all the time. The static landscape image is replicated by Chinese landscape architects just like the prototype of picturing landscape exist in the time of Olmsted. In this sense, it is supposed to develop a more dynamic landscape conception in combination with the evolving ways of spatial organization under the framework of the shan-shui structure.

In this book, learning their critical rationalism approaches to urban landscape analyses is worthwhile in the face of Western theoretical and practical experiences on *landscape urbanism* and *landscape structuralism*. For China, blind replication and the lack of a critical approach have often failed to offer an authentic way out. Moreover, these issues are not the main purpose of this research. This study will answer the question: what should be thought of in the process of referencing? Hereto, German cultural theorist Hartmut Böhmethus seems to give us a suggestion from a cultural perspective. He pointed out that our demands are the "establishment of cultural reflection" in the societies themselves (Böhmethus 2000). Such a cultural reflection indicates that theoretical analysis and comprehension of the urban landscape and shan-shui parks should generally be directed toward cultural identity formation based on a rational, critical reference to developed regions within the Chinese socio-cultural context.

In this sense, the further conception of Chinese shan-shui parks can be in complete accordance with neither the North American organic nor the German *structuralistic approach*. Instead, crucial points discovered from North American and German park design paradigms perhaps call for the ongoing self-development of the distinctive and diverse Chinese contemporary large-scale park approaches in the future.

Based on the organic approach, some thought-provoking ecological ideas of nature are valuable to the conception of shan-shui parks. Specifically, an increase in ecological awareness does not stimulate the formation of diverse ecological ideas in the professional field with the major environmental challenges faced by most Chinese cities. The Chinese urban landscape has been devoid of ecological theoretical support, whereas the North American urban landscape has been theoretically implemented creatively, where landscape and ecology are conceived as "agents of creativity" (Corner 1997). Through systematic analysis, emerging ecological thoughts that have emerged since the 1980s and infused into North American organic parks are revealed in Chapter 4. Among these thoughts, the deduced characteristics of complexity and resilience from the landscape-ecological perspective articulated may be considered in future shan-shui park conceptions.

Referring to the *structuralistic approach*, the post-industrial site is reframed and expressed through the structure, in which almost everything is retained. The site is seemingly inherited and based on durable development over time. The German approach implies a coherent landscape understanding of the specific cultural contextualization. This will be clearly illustrated in the comparative part of the Chapter 5. Similarly, the inherent structure is considered in Chinese traditional planning and design. Such a consideration manifests as the shan-shui structure that reflects the traditional shan-shui culture within the cosmology of harmony between nature and man. The future development of shan-shui parks is inseparable from the discussion and rethinking of spatial structure.

In this sense, the constant self-development of the Chinese shan-shui park approach still needs time to adapt to the contemporary Chinese urban environment. The representation of the spatial structure can either be the traditional shan-shui structure or its abstraction with more individual creativity. To promote the new development of these large-scale urban parks, it is essential to ensure the compatibility of contemporary urban landscape analyses and formulations with shan-shui parks.

In conclusion, through the critical rationalism approach, contemporary urban landscapes are therefore on the way to a readjustment, particularly in developed regions. It drives landscape architects to foster a "critical" and "professional" understanding of landscape at an urban level. The two essential strands were seized by Waldheim in 2006. In the next section, the urban landscape readjustment in North America and Germany is described in a dynamic urban environment.

# 3. Contemporary Urban Landscape Conceptions

"The City of the future will be an infinite series of landscapes: psychological and physical, urban and rural, flowing apart and together [ … ]. Christopher Alexander was right: a city is not a tree. It is a landscape." (Turner 1996)

## 3.1. Contemporary Urban Landscapes in Adjustment

This book highlights the renewed understanding of how three large-scale urban park concepts are influenced by contemporary urban landscapes. It is gained by readjustment, which definitely involves a critical attitude. The implication is that the academic circles in both North America and Germany began to advance in their own understanding of the term "landscape" at the urban level. Similarly, China is also making attempts to explore the significance of landscape at the same level from the perspective of urban regeneration. In this part, the tendency of readjustment is reflected mainly in the preparation made for the further formulations of *landscape urbanism* and *landscape structuralism*. This may shed some light on the conceptual path to urban landscape for China in the context of cultural coherence.

In the meantime, the readjustment of the urban landscape is bound up inextricably in the revised city, that is, dissolved urban structure and transitional urban society. It is argued in this section that the ongoing changes to the urban environment contribute new perspectives to viewing and interpreting contemporary urban landscapes and their essential embodiment as large-scale urban parks in a critical way (De Jong 2000).

### 3.1.1. A New Landscape at the Urban Level

In recent years, as Charles Waldheim's clear-cut declaration on "realignment" (Waldheim 2006) in North American landscape architecture outlines, much more information on the readjustment of the term "landscape" seems to have been captured by worldwide landscape architects. However, "large-scale landscape architecture shifting into a planning discipline" has become a tendency, revealing progress in the concept of landscape both in countries in North America and Germany (Schöbel and Czechowski 2009). Thus, the term "landscape" assumes new delineations at the urban level that not only broaden its connotation but also make its role in urban regeneration prominent.

For example, the derelict infrastructure of the railway in the Meatpacking District in Manhattan was transformed by the prominent High Line on a post-industrial site in New York designed by James Corner and Field Operations into a linear urban park as "a new paradigm for the promenade" (Dümpelmann 2018). The project presents a new landscape that has emerged according to the renewed comprehension of urbanity, urban nature, and regional identity in the context of urban renewal, as well as the concept of "city as landscape" as put forward by English landscape architect Tom Turner, as shown in Figure 16.

**Figure 16.** High Line in New York creating a new journey for visitors worldwide, Diller Scofidio + Renfro and Field Operations, 2009. Source: Photo by ©Zhenkun Gan 2019; used with permission.

To account for the emergence of the tendency in landscape architecture, James Corner presented three aspects as key factors: the remarkable increase in global ecological awareness, the need for regions to retain a sense of unique identity, and the impacts of massive urban growth on rural areas (Corner 2006). In other words, the promotion of ecological sustainability in search of regional identity and the rise of rural areas stimulate the new landscape articulations in developed regions. With the integration of urban natural, cultural, and spatial factors into the critical discussion on landscape, the new landscape at the urban level is formed in different ways in North America and Germany.

In North American academe, landscape architects, such as Charles Waldheim, James Corner, and Elizabeth K. Meyer, advocate readjustment ambitiously. This advocacy is mostly reflected in matters related to landscape's significance and the expansion of its "scope" and "scale" (Corner 1999). Given these three areas of significance, "scope", and "scale", James Corner's articulated term for *landscape* holds central significance within design professions, such as architecture, landscape architecture, and urban design and planning.

Moreover, there is a shifting "interest in a deep concern with landscape's conceptual scope; with its capacity to theorize sites, territories, ecosystems, networks, and infrastructures, and to organize large urban field" (Corner 2006). The landscape "scale" extends naturally to metropolitan areas as the conceptual consideration of landscape capacity at the urban level is identified. For example, Queen Elizabeth Olympic Park constructed by Hargreaves Associates on a post-industrial site in London as a large-scale environmental restorative network demonstrates the possibility of landscape to change the appearance of a city and drive the re-development of urban areas, as shown in Figure 17.

**Figure 17.** Queen Elizabeth Olympic Park weaving the spatial fabric of London, Hargreaves Associates, 2012. Source: Photo by author.

Another orientation of readjustment is placed on the combination of landscape and ecology, when North American landscape architects attach more importance to the role of 'ecological awareness', which is the aforementioned first factor proposed by James Corner. These professionals are acutely aware of certain crucial changes touching upon dynamic ecosystems with renewed characteristics, and then reintegrate their understanding of nature with urban landscape planning and design. Undoubtedly, they offered overwhelming support for ecological sustainable development. Hereto, Nina-Marie Lister purported the ecological impact upon large-scale landscape architecture as "the renaissance of landscape" throughout the last fifteen years (Lister 2015). Increasing landscape architects, who are guided by ideas and principles in ecological sciences, have affirmed the development of urban landscape "coupled with a focus on indeterminacy and ecological processes as catalysts for the reemergence of landscape theory and praxis" (ibid.). Consequently, the readjustment of landscape in North American landscape architecture contributes to the formation of the concept of *landscape urbanism* in the mid-1990s, which will be articulated in the following part.

In German academe, readjustment is also viewed as "the renaissance of landscape because of the rise of suburbia" (Schöbel and Czechowski 2009). Sören Schöbel explained:

"The dissolution of the evident distinction between city and landscape, between urban and rural areas, is leading to a development of a new form of city and a new form of landscape not only in terms of building and infrastructure but also concerning lifestyle and social relations. Specific urban landscapes appeared." (ibid.)

As one of rational analyses in German landscape architecture, Sören Schöbel's statement expressed the consideration of new landscape at the urban level, claimed as *critical reconstruction* (German: *Kritische Rekonstruktion*) that was known from *urban design* (German: *Städtebau*) in the 1980s (Schöbel 2014). Sören Schöbel proposed that modern urban planning was replaced by *careful urban renewal* (German: *Behutsame Stadterneuerung*), *dialogical urban development* (German: *Dialogische Stadtenwicklung*), and critical reconstruction of urban texture (German: *Kritische Rekonstruktion der Stadttextur*) (ibid.).

Early in the 1960s, there was a critique of the mechanically conceptual model of the modern city with pre-planned functional zoning and distribution in urban planning. Among them, Italian architect Aldo Rossi, in his work *The Architecture of the City*, argued that the true essence of a city is deprived in the architectural practice (Rossi 1984). The city should be understood and valued as a physical and social arrangement constructed over time. Aldo Rossi's view actually laid a foundation for the more anti-modernist ideas. In the 1970s, sociologist Richard Sennett's view of the careful rebuilding of European cities was confirmed by all empirical findings in the historical context on social, economic, and ecological benefits. In other words, the development of a city connected to the historical past would have needed to be constructed carefully over time.

Apparently, an increasing number of architects and urban planners at that time were critically aware of modernism itself owing to the "emptiness and dissatisfaction they felt in the urban environment" (Barrows 2010). They even attributed the destruction of the city less to the Second World War than to the functionalistic idea held by professionals in urban planning (Lampugnani 1983). As a result, they adopted a scientific critical approach central to their empirical practices.

German architect Josef Paul Kleihues, who contributed to the *critical reconstruction* of Berlin from 1984 to 1987, aspired to return to traditional urbanism. Particularly, Kleihues applied his own concept of *critical reconstruction* to urban renewal projects of IBA in the 1990s, in which he advanced traditional urbanism, highlighted the mixture and integration of urban functions, and shaped the overall "character" of a city and "differentiated architectural forms" (Kleihues and Rathgeber 1993). Given the critique and restraint on modern functionalist ideas as regards European cities, the *critical reconstruction* program emerged in the field of German urban planning. In the course of *critical reconstruction* in Berlin, Potsdamer Platz and its spatial structures were redesigned for the commercial redevelopment of large plots. This is premised on fully inheriting the morphological characteristics of the original small neighbourhoods of the historic city. For instance, according to the street scale and spatial texture, a park of the Prachtgleis named Tilla-Durieux-Park in the new building complex was also created to become a large green space amidst the urban density of the Potsdamer Platz sector. This park, a turf sculpture, provides the center of Berlin with a popular recreational space. It also helps people to realize the virtues of the empty space that has characterized the area for many years, as illustrated in Figure 18.

**Figure 18.** Tilla-Durieux-Park in Berlin, DS Landschapsarchitecten, 2003. Source: Photo by author.

Taking such history into account, the concept of *critical reconstruction* was adopted in the field of German landscape architecture to rethink the contemporary urban landscape. The transition of this concept is possible, arising from an implicit association of the 1980s *critical reconstruction* with the urban landscape, owing to the identical anti-modernist perspective. This point was affirmed by Italian architect Vittorio Magnago Lampugnani's statement:

"The European and especially the West German urban landscape has clearly been destroyed less by the war than by the planners who, because of their abstract, biased and global conception of a city which in their view is an addition of quantitative functions, have turned them mostly into cheerless and desolate places." (Lampugnani 1983)

Meanwhile, the adoption of the critical reconstruction concept was also noted in Sören Schöbel's 2014 essay Landschaft–kritische Rekonstruktion through a rhetorical question:

"Lässt sich auch Landschaft in einer 'kritischen Rekonstruktion' entwickeln? Dazu ist es erforderlich, auch in der Landschaft jene Elemente, Bausteine und Typologien zu identifizieren, die eine gewisse Stabilität über die Zeit hinweg aufweisen können und dabei in der Lage sind, Vielfalt in einem Zusammenhang zu fördern." (Schöbel 2014)

Sören Schöbel's question stressed the necessary requisite of developing *critical reconstruction* in German landscape architecture. In the urban landscape, its certain stability could be presented over time, with the identified elements, building blocks, and typologies in landscape. Thus, "diversity and differences" of urban landscape "as immanent qualities" can be promoted (Schöbel and Czechowski 2009). In this sense, the urban landscape is dependent on spatial qualities through specific formal elements, rather than merely concentrating on functions. These elements, building blocks, and typologies are deployed for the characteristic urban fabric, instead of homogenous urban texture. They are also organized into "a spatial structure, an open entireness where diversity and differences are and where coherence could be generated" (ibid.). Specifically, urban landscape is visualized and established by a structure composed of multilayered landscape elements and typologies, in which 'diversity', 'differences', and 'coherence', as essential qualities, consider both historical and current contexts.

For instance, there is a cluster of historical industrial structures, the warped-looking buildings designed by American architect and designer Frank Gehry and other postmodernist buildings in the Media Harbour located on the river Rhine of Düsseldorf's docks. Through the elements of the river, bridge, plaza, buildings, infrastructure and remaining industrial structures, they built a diverse yet coherent urban landscape for revitalizing the city, as illustrated in Figure 19.

**Figure 19.** Media Harbour in Düsseldorf, Frank O. Gehry, David Chipperfield, Joe Coenen et al., 1990. Source: Photo by author.

In North America and Germany, their respective readjustment leads the urban landscape to be redefined in two different yet similar new ways. On the one hand, as the significance, 'scope', and 'scale' of the landscape at the urban level promoted, the North American term *landscape* is adjusted along the eco-priority track to cultivate a resilient urban landscape according to an ecosystem property of resilience that will be explained in Canadian ecologist C. S. Holling's 1992 dynamic model in Chapter 4. In this situation, the *landscape urbanism* program rose with the critical approach, regarded as "a robust alternative to the failures of modernist urban planning" (Tully 2013). This part is analyzed in the exposition of *landscape urbanism* with an organic approach. Its proponents are precisely the scholars who support landscape readjustment actively.

On the other hand, under the influence of the *critical reconstruction* program in urban planning, the German term *landscape* is adjusted on the way to the structure for shaping a characteristic urban landscape with regional cultural landscape elements, presenting spatial "diversity", "differences", and "coherence". Accordingly, German *landscape structuralism* with a *structuralistic approach* in research is also further discussed in parallel with North American *landscape urbanism*.

In this book, we also trace back to the physically changing urban environment that invites new ways of seeing and interpreting different urban landscapes before the two kinds of urban landscapes in developed regions are stated. It embraces the shifts in contemporary urban structure and urban society, involving two aspects of the research question. Then, essential theoretical analyses of urban landscapes are elicited for better consideration of two theoretical schools of thought: *landscape urbanism* and *landscape structuralism*. They constitute the following main sections.

## 3.1.2. Diverse Analyses of Urban Landscapes

In the field of landscape architecture, there are more than two kinds of conceptual recognition of contemporary urban landscapes that incline toward certain organisms and *structuralism* in North America and Germany, respectively. However, the critical approach helps intellectually distinguish two ranges of different concepts of urban landscapes: *landscape urbanism* and *landscape structuralism*. They are two different theoretical schools of urban landscapes with their own conceptual approaches, ideas, and focuses. These two categories are selected owing to their intimate connections with current large-scale urban park concepts and actual projects. In other words, the two urban landscape frameworks largely support the planning and implementation of many large-scale urban park projects in North America and Germany, such as Freshkills Park and Landscape Park Duisburg-Nord. Urban landscapes in these two schools reflect that their critical analyses recommend identifying not only the characteristics of new and twenty-first-century landscapes but also the manner by which they are conceived in regional, socio-cultural, economic, and ecological conditions.

Especially notable is the meaning of *school* that could be found in the term *school of thought*; such meaning is "a point of view held by a particular group" in the American Heritage Dictionary definition. The North American urban landscape theoretical schools of thought are led by "people who actively write about the theories of landscape urbanism [ . . . ]: James Corner, Stan Allen, Alex Wall, Charles Waldheim" (Duncan and Seltzer 2010). James Corner is among the "thought leaders" (ibid.) who are closley associated with concepts and approaches of North American urban landscape and large-scale urban parks. This is in that he had more influence in the book than the others, and particularly, his *critical thinking* approach and ecological ideas involved the comprehension of *landscape*. Consequently, the so-called James Corner *school of thought* is emphasized in this work's explanations.

Accordingly, *structuralism* is essentially "a school of thought initiated in the early twentieth century by the great linguist Ferdinand de Saussure," according to the explanation of the Dream Encyclopedia (Lewis and Oliver 2009). *Structuralism* has been developed from its origin and influenced by the Dutch movement of architectural *structuralism* since the 1960s (Peisl 2014), for instance, the Cube Houses designed by Dutch architect Piet Blom in central Rotterdam are among the representatives of this *structuralism* movement, as shown in Figure 20. Peter Latz is among the German landscape architects who combined *structuralism*'s theoretical parts, such as "the writings of architects like Aldo van Eyck and Herman Hertzberger" (Weilacher 2008), to expand its meaning in German landscape architecture. In this sense, Peter Latz plays a key role and is also considered one of the thought leaders. Hence, the so-called Peter Latz *school of thought* in this book is applied to the further comprehension of German urban landscape and its structuralistic parks.

**Figure 20.** Cube Houses in Rotterdam, Piet Blom, 1977. Source: Photo by author.

In-depth discussions on urban landscape formulations are conducted based on two sequential stages of urban landscape analyses in North America and Germany. The first analytical stage involves theoretical foundations from the 1970s to the 1980s; the second analytical stage concerns specific theoretical orientations in the 1990s. The transition from theoretical analyses to formulations is presented in Tables A1 and A2. Certain key information is also concluded in Tables A3 and A4. James Corner's and Charles Walderheim's ideas contribute to North American urban landscapes and are influenced by J. B. Jackson's 1984 "vernacular-mobile" landscape understanding. Peter Latz's and André Corboz's views play an essential role in the German urban landscape concept and are guided by Henri Lefèbvre's 1974 analysis of "social production of space". As aforementioned, James Corner and Peter Latz are representative personalities of the two theoretical *schools of thought* owing to their leading critical rationalism approaches of *critical thinking* and *critical structuralism* and conceptual organic and structuralistic approaches to urban landscapes.

Stage I: Theoretical Foundations of Urban Landscapes

From the 1970s to the 1980s, initial analyses on urban landscapes laid a solid foundation for further exposition of urban landscapes in Europe and North America. Henri Lefèbvre and J. B. Jackson dedicated several studies to the conductiion of such analyses.

From a sociological perspective in Europe, the comprehension of the urban landscape was bolstered at the beginning of the 1970s by Henri Lefèbvre in his book *The Urban Revolution*. The sociological influence on urban space stems from German sociologist, philosopher, and critic Georg Simmel's idea at the turn of the twentieth century. Georg Simmel claimed that "the city is not a spatial entity with social consequences, but a sociological entity that is formed spatially" (Simmel 2007). His idea of urban sociology may play a significant role in Henri Lefèbvre's social organization of space that would be interpreted subsequently. Additionally, Henri Lefèbvre's urban landscape analysis was performed based on a hypothesis of complete urbanization of the world (Smith [1970] 2003). On account of the widespread urbanity, the urban landscape would become a global proposition to be discussed continuously and widely.

Moreover, Henri Lefèbvre's research on the urban landscape could be summarized in two aspects. In the first aspect, the European urban landscape is more defined within a scope of "mixed or intermediating" "level M", that is, the "specifically urban level" compared with dimensions of 'superstructure' and 'infrastructure' (Lefèbvre [1970] 2003). Specifically, the urban landscape on this level is not only considered "green infrastructure" measured as functions and interpreted as metaphors but also analyzed as social and spatial forms of nature (Schöbel and Czechowski 2015), as shown in Figure 21. This cognition elicits the other analytical aspect of Henri Lefèbvre research as follows.

**Figure 21.** The sustainability of green open space stemming from the diversity of urban life, Munich. Source: Photo by author.

The second aspect concerns Henri Lefèbvre's crucial perspective of "social production of space" in 1974. He described that "space is produced and reproduced through human activity and it thus represents a site of struggle and contestation. It is not an empty container simply waiting to be filled" (Lefèbvre [1974] 1991). "Space and the political organization of space express social relationships but also react back upon them" (ibid.), as demonstrated in Figure 22. Through the critique of this idea, increasing geographers, sociologists, and cultural scientists realized that production conditions and social awareness are structuring factors of not only society but also space (Schöbel and Czechowski 2015). In other words, this idea urged many European researchers to discover the role of space in the constitution of social relationships (ibid.). Therefore, Henri Lefèbvre's understanding of social space in everyday life triggered an essential movement in space-related academic research, namely, spatial turn.

However, the significance of identifying social space lies in guiding the urban landscape into the level of "difference" with the ubiquity of urbanization and globalization (Lefèbvre [1974] 1991). This point carries huge implications for German urban landscapes to a considerable extent. For Henri Lefèbvre, "(social) space is a (social) product" (ibid.). Thus, the 'difference' in every society is unfolded through every distinct mode of production that produces a certain space of its own society. The interaction between urban space and complex social construction leads to the 'difference'. According to Geman scholar Stefan Körner, urban landscape in specific socio-cultural contexts could be shaped intendedly and unintendedly as every-day-use-related landscape design; this design might no longer be associated with Arcadian harmonies on the super level while pointing to *Eigenart* in German, which indicates the concept of the character of a culture or space that is signified by the term *Eigenart*, and hence, is full of character (Körner 2013). The focus on 'difference' in the early stage of urban landscape analysis contributes to the theoretical orientation of German urban landscapes during the 1990s, which is also explained in the following part.

**Figure 22.** The urban space constantly organized through social relations, Dubrovnik. Source: Photo by author.

Besides Henri Lefèbvre's analysis and influence in Europe, another important scholar during the early 1980s was J. B. Jackson, whose contribution to urban landscapes in both North America and Germany is rebuilding a modernized understanding of the term *landscape* (Höfer and Trepl 2010). J. B. Jackson's analysis is uncovered continuously by North American and European scholars when they tend to explore the essence of landscape: whether "an ideal aesthetic construct, a physical place of human interaction, or both ideal and object" (ibid.). For instance, regarded as the vital model for one type of urban park reconversion of disused industrial sites, Gas Works Park in Seattle designed by American landscape architect Richard Haag manifests an art of nature with the thought of 'Zen' and a place for human leisure activities, as shown in Figure 23.

**Figure 23.** Gas Works Park in Seattle, Richard Haag, 1971. Source: Photo by ©Shuang Zhao 2017; used with permission.

Moreover, through Richard Haag's park design practice, the development path of post-industrial areas is renewed, and the interdisciplinary knowledge is applied to the decontamination and restoration of abandoned land for the first time, showing the landscape architects' preliminary explorations of large-scale urban parks. He gathered various professionals who brought together a series of different skills to verify unconventional techniques for solving the soil-pollution problem. Working with the chemical engineer Richard Brooks, he came up with the solution of removing the contaminants from the surface by activating the indigenous bacteria by tilling sawdust, sewage sludge, and other organic compounds. As one of the first examples of bio-remediation, the process proves that contaminated areas could be safely reclaimed. Gas Works Park showcases a certain form of organic thinking and represents the prototype of former industrial sites reconverted as urban parks in North America.

Given the new understanding of the term *landscape*, for J. B. Jackson, the repositioning of *landscape* in its contemporary meaning boils down to the "vernacular-mobile" landscape proposed in his 1984 book *Discovering the Vernacular Landscape*. In a broad sense, his consideration is based on a cultural perspective. Against socio-cultural backgrounds, the definition of *landscape* varies as specific territories. This aspect stimulated an increasing number of landscape architects to actively explore the unique relationship between social and spatial changes over time and the creation of their own organizations of spaces. For instance, Tanner Springs Park in Portland, USA, is transformed over time for the purpose of reshaping its wetland origins on the industrial wasteland. In this area, the approach of designed ecology is adopted to organize a unique, local space with the characteristics of art, artificial nature and mixed human uses, as illustrated in Figure 24.

**Figure 24.** Tanner Springs Park in Portland, open to social and ecological processes over time, Ramboll Studio Dreiseitl, 2005. Source: Photo by ©Shuang Zhao 2017; used with permission.

For J. B. Jackson, the "vernacular landscape" could be conceived in a distinct way to "define and handle time and space" (Jackson 1984). The key to this distinct way lies in realizing a juxtaposition of reality and idealisation, that is, "mobility" and "permanence", in the concept of landscape (ibid.). He deliberated their correlation and posited that the contemporary landscape is not always in a state of permanence:

"A landscape, like a language, is the field of perpetual conflict and compromise between what is established by authority and what the vernacular insists upon preferring [ … ]. Whatever definition of landscape we finally reach, to be serviceable it will have to take into account the ceaseless interaction between the ephemeral, the mobile, the vernacular on the one hand, and the authority of legally established, premeditated permanent forms on the other." (Jackson 1984)

In contrast, the reality, one's everyday world, in a contemporary urban landscape, concluded as the everyday landscape, demonstrates landscape "in mundane terms" (ibid.). One could find identity from their daily lives, leading one to see critically a "landscape as something more than beautiful scenery" (ibid.). As mentioned by J. B. Jackson, the everyday world appropriated by various kinds of individuals may form the everyday landscape subject to temporary mobility and change. The temporary mobility and change offer a possibility of local self- determinacy or self-organization by various ordinary people under the idea of social equality and liberty. The self-determinacy indicates that people change landscape gestalt without a predetermined purpose in the process of living on the land temporarily (Höfer and Trepl 2010). In an ideal and long-standing cognition, landscape is "a vista or view of scenery of the land," "a work of art," and "a kind of supergarden" (Jackson 1984) since the perception of landscape is derived from tangible nature and the "organic unit of organic society" (Höfer and Trepl 2010). This is an ideal social order.

In J. B. Jackson's landscape concept, thus, there is a co-existence of "Landscape Two," "a landscape identified with a static, very conservative social order and that there can be only one true philosophy of nature" (Jackson 1984), and "Landscape Three," a "dynamic system of manmade spaces", elaborated as follows:

> "Landscape is not scenery, it is not a political unit; it is really no more than a collection, a system of man-made spaces on the surface of the earth. Whatever its shape or size, it is never simply a natural space, a feature of the natural environment; it is always artificial, always synthetic, always subject to sudden or unpredictable change. We create them and need them because every landscape is the place where we establish our own human organization of space and time. It is where the slow, natural processes of growth and maturity and decay are deliberately set aside and history is substituted. A landscape is where we 'speed up' or retard or divert the cosmic program and impose our own." (ibid.)

Moreover, the co-existence of these two kinds of landscape concepts may lead to a "dilemma" (Prominski 2010). This situation, indicated by German landscape architect Martin Prominski, informs landscape architects not to fall into a simplistic perspective, that is, "either one or the other" (Beck 2008). In spite of the "dilemma", it is necessary to use a critical perspective with respect to the definition and conception of landscape, as demonstrated by J. B. Jackson's analyses of "Landscape Two" and "Landscape Three". Ultimately, North American landscape architecture, which adopted the ideas of J. B. Jackson, shifted its attention from the ideal and permanent landscape with harmonious, beautiful, and natural characteristics to a mundane and everyday landscape with realistic, dynamic, and unpredictable urban characteristics.

On the basis of the aforementioned urban landscape analyses in North America and Europe, the second analytical stage provides that theoretical orientations have unfolded since the 1990s because urban landscape concepts "do not only describe realities, but suggest orientations" (Wolfrum and Schöbel 2011). The orientation is considered either a "metaphor" (Corner 2006) in the North American analysis or a "theoretical construct" (Ipsen and Weichler 2005) in the German analysis.

Stage II: The North American Analysis: Metaphor

Since the 1990s, North American urban landscape analysis has reflected metaphor as orientation. James Corner epitomized the landscape, which affords a range of "imaginative and metaphorical associations" (Corner 2006). Metaphor refers to a metaphorical conceptualization of cities through the "lens" of landscape stated by Charles Waldheim in his 2006 work *A Reference Manifesto*, with cultural embedding of imagination. In other words, landscape is regarded as a conceptualized model of describing and envisioning contemporary cities, as shown in Figure 25. Metaphor implies an essential shift in understanding North American cities from the perspective of landscape, which has been summed up in a hypothesis of "landscape as urbanism" (Waldheim 2006). This shift verifies Dutch architect Rem Koolhaas's 1998 definition of landscape as the "primarily element of urban order" and Charles Waldheim's 2006 definition as the "medium" to construct a city. It also helps the meaning of landscape recover from "a framed static picture to acting as operational and performative" (Assargård 2011). For this static and scenic image of the landscape, James Corner criticized that "landscape is nothing more than an empty sign, a dead event, a deeply aestheticized experience that holds neither portent nor promise of a future" (Corner 1999).

**Figure 25.** Landscape planning scheme of regional Beijing-Zhangjiakou Railway Heritage Park, China Architecture Design & Research Group, 2019. Source: Photo courtesy of ©China Architecture Design & Research Group; used with permission.

Moreover, the analogy between city and landscape is further drawn by the sciences of ecology, which is an indispensable potential factor. Consequently, "the city is like a landscape" and "will function like a landscape", as indicted by the ecological metaphor (Tully 2013). American designer and researcher Chris Reed mentioned the influence of the sciences of ecology that "the past two decades have witnessed a resurgence of ecological ideas and ecological thinking in discussions of urbanism, society, culture and design" (Reed and Lister 2014). They also articulated a tendency in ecological sciences that have moved away from "classical determinism and a reductionist Newtonian concern with stability, certainty and order, in favor of more contemporary understandings of dynamic systemic change and the related phenomena of adaptability, resilience and flexibility" (ibid.). These concepts in the critical cognition of ecology are viewed as "models or metaphors for cultural production" (ibid.).

In this work, metaphor could be understood properly through ecology or precisely dynamic, fluid, complex, and indeterminate ecosystems known by both ecologists and landscape architects particularly beginning in the 1980s. This understanding results from scientific studies and discoveries as regards dynamic ecosystems and, subsequently, newly emerging ecological ideas that have entered into the field of landscape architecture in the wake of the *landscape urbanism* program. This suggests ecological changes in the ecosystem paradigm and relevant innovative views on nature. The conceptual changes caused by ecology are related intimately to the mid-1990s *landscape urbanism* and its implementation of North American organic parks. These specific ecological changes will be detailed in Chapter 4.

To proceed from the point of ecological metaphor, North American landscape architects tended to imagine the city as a fluid living organism before James Corner presented the idea of "a more organic, fluid urbanism" (Corner 2006). In this sense, city, landscape, and ecology are considered in an integrated approach. The metaphor also becomes the key to analyzing North American urban landscape and ecology, which will be deduced as a comparative aspect in Chapter 5, accompanied by landscape understanding and landscape and life.

In North American landscape architecture, the metaphor of ecology is manifested in the philosophy of "interconnection and codependency between organisms and environments, between objects and fields" (Weller 2006). All things are interconnected to each other on the extensive urban surface. Consequently, Australian landscape architect Richard Weller asserted that "the city in mind here is not a place or just 'a' system, but a part of all processes and systems, a field which covers and makes up the world at any given time" (ibid.). Guided by the philosophy, proponents of *landscape urbanism* are concerned with the ecological metaphor and thus prioritize the relationships between things over objects alone.

Generally, the metaphor becomes among the most distinguishing features of *landscape urbanism* and has exerted a potential influence on the North American organic park concept. Julia Czerniak and George Hargreaves collectively and clearly pointed out that the 2007 book *Large Parks* following Charles Waldheim's 2006 work *The Landscape Urbanism Reader* is another key direction to promote the exploration of North American urban landscape progress. In other words, *landscape urbanism* is inextricably linked to our research proposition of large-scale urban parks, with an organic approach that will be systematically argued in Chapter 3.

Stage III: The German Analysis: Theoretical Construct

The term *urban landscape* is mentioned as a category of space in German analysis in the early 1990s. Sören Schöbel's point of view is recognized as a relatively young but (widespread) common technical term, and used by summing up the following various phenomena well known in the professional field, such as *suburban area, Zwischenstadt, city landscape, city region, sprawl, periphery, commuter belt*, and *urbanization* (Schöbel and Czechowski 2013).

Regarding the phenomena of urban spaces, the German understanding of the urban landscape possesses its own theoretical orientation. "As a core concept for inquiry into these new urban spaces," Detlev Ipsen and Holger Weichler propose the term *urban landscape* (Ipsen and Weichler 2005). It is a term that one does "not understand as a metaphor, but rather as theoretical construct that opens up an interdisciplinary path for the analysis and planning of urban regions" (ibid.). The "interdisciplinary path", in Sören Schöbel's explanation, implies "various space-describing and space planning disciplines, such as geography, sociology of space, urban studies / urban development, architecture, and landscape architecture" (Schöbel and Czechowski 2013).

Specifically, two aspects support the urban landscape as a theoretical construct. According to Sören Schöbel's opinion, they could be summarized into the urban landscape (which is essentially associated with urbanity) and urban landscape (with a change in urban spatial structure), expressed as follows:

"On the one hand, urban landscape describes the complete urbanization of space analytically (i.e., the overall expansion of urban designs, infrastructure and lifestyles). On the other hand, it programmatically describes experiments to detect and design new relations in fragmented areas which are neither city nor country." (ibid.)

Notably, underlining the "difference" within the ubiquitous urbanity becomes a core in the process of theoretical construction. European scholars, such as Henri Lefèbvre ([1974] 1991), Thomas Sieverts (2008), and Sören Schöbel (2015), have explicitly proposed the "difference". The essence of "difference" could be traced back to Henri Lefèbvre's *social production of space*, which has been analyzed above.

Furthermore, Thomas Sieverts improved their quality as one central point in his view of "fragmented urban landscapes" in 2008: "urban landscapes as new forms of urbanity can only become productive if they can develop their own particular characteristics, leading to productive distinctions in economy and culture" (Sieverts 2008). He stated that with regard to the area's own distinct characteristic, the 'difference' must be the first element of design and, thus, there is "the need for quality improvement" (ibid.).

Additionally, Sören Schöbel supported the "difference" by treating urban landscapes as "specifically describable landscapes," rather than as "featureless" or "generic" areas (Schöbel and Czechowski 2013). The primary reason for such treatment is that urbanization and globalization are presumed to not lead to indistinguishable and generic cities but reinforce the "difference" through which landscape as specific forms of urbanity could be developed (Schöbel et al. 2013).

With the urban landscape analyses in two stages of different theoretical foundations and orientations in North America and Europe, the formulations of *landscape urbanism* and *landscape structuralism* in different cultural contexts could be further analyzed in the following part.

## 3.2. Landscape Urbanism in North America

In the mid-1990s, the emerging notion of *landscape urbanism* was an initiative born in North America (Thompson 2012). Two relatively immediate factors play a part in its emergence. Above all, its supporters, who searched the theoretical framework in the writings of early regional planners, including British biologist and town planner Patrick Geddes, American forester and planner Benton MacKaye, American historian, sociologist and philosopher Lewis Mumford, and particularly Ian McHarg, recognized that *landscape urbanism* could benefit directly from the canonical texts of regional environmental planning (Waldheim 2006).

Moreover, *landscape urbanism* is regarded as "a robust alternative to the failures of modernist urban planning" (Tully 2013). The origin of *landscape urbanism* could be traced to postmodern critiques of modernist architecture and planning; the early critiques were derived from the field of architecture as early as the 1980s and then expanded to the field of landscape architecture (Waldheim 2006). Charles Waldheim, one of the staunch advocates of *landscape urbanism*, emphasized its strength that "it offers an implicit critique of architecture and urban design's inability to offer coherent, competent, and convincing explanations of contemporary urban conditions" (ibid.).

The concept of *landscape* is defined as having a focus on process and systems philosophy instead of the former focus on pastoral images as the *landscape urbanism* program emerged with a critical attitude (Reed and Lister 2014). "A hallmark of landscape urbanism is the understanding of ecological systems and the knowing of processes that constitute them" (Gray 2006). As discussed in the ecological metaphor, the newly defined concept of landscape is fostered owing to the intersection of ecological sciences and landscape architecture. In the *landscape urbanism* framework, an organic approach is supported by the understanding of the role of ecological sciences.

*Landscape urbanism* certainly draws upon "terms", "conceptual categories", and operates methodologies of field ecology for the understanding of site and city (Waldheim 2006). The terms, such as "diversification, flows, complexity, instability, indeterminacy, and self-organization, become influential design generators, shaping the way we consider and construct places" (Corner 1997). These terms broaden the horizons of landscape architects to analyze and highlight the occurrence of spaces and spatial *performance*, implying the effectiveness of their conceptual urban field with permanent fluidity and adaptation. These terms also represent an organic method to shape spatial processes and interpret natural systems.

Meanwhile, "conceptual categories" as "movement diagrams" are also employed actively from the landscape-ecological perspective (Reed and Lister 2014). These categories are developed by Richard T. T. Forman based on his ecological research on "the availability of LandSat imagery and computer-aided geographic information systems analysis during the 1980s and early 1990s" (ibid.). The conceptual categories are generally recognized as patches, edges, corridors, mosaics, and matrices, which will be used mostly by North American landscape architects to establish an overall conceptual diagram for concrete urban landscapes, such as North American large-scale urban parks. These will be illustrated by their project cases.

These conceptual categories become essential dynamic patterns to understand ecosystems described as matrices and networks and characterized by adjacencies, overlaps, and juxtapositions (Forman et al. 1996). With landscape-ecological research on ecosystems, these categories' dynamic living nature not only embraces physical elements but also supports the movement and exchange of substances with changing conditions. For example, the Olympic Sculpture Park located in Seattle as a new green space and dynamic ecosystem is re-developed on a former polluted industrial site. Featuring the landscape progression and interaction from upland to shoreline through the distinct valley, grove, meadows, and shore, it re-introduces the complexity of habitat to the site by restoring the original topography and utilizing native plants, as illustrated in Figure 26. Meanwhile, it is also increasingly accepted by landscape urbanists. Moreover, the processes of redefining the conceptual categories and discovering the terms imply a radical paradigm shift of ecosystems from equilibrium to non-equilibrium. These processes and their resulting paradigm shift will be explored in Chapter 4, covering North American organic parks with newly emerging ecological ideas.

**Figure 26.** Olympic Sculpture Park in Seattle, Weiss/Manfredi Architects and Charles Anderson Landscape Architecture, 2007. Source: Photo by ©Shuang Zhao 2017; used with permission.

*Landscape urbanism* absorbs the terms and conceptual categories in the ecological field to build up its own organic approach to conceive urban landscapes. One of the most active advocates of this approach in the field of landscape architecture is James Corner, who emphasised the creative potential of ecology in his 1997 essay, titled Ecology and Landscape as Agents of Creativity. His influence of ecological ideas on urban landscapes also reveals why the James Corner school of thought is identified in the book as an urban landscape school of thought. James Corner, with *critical thinking*, claimed "a creative relationship with ecology for exploiting a potential that might inform more meaningful and imaginative cultural practices than the merely ameliorative, compensatory, aesthetic, or commodity oriented" (Corner 1997). In his *cultural imagination*, landscape is defined as "innovative cultural agent" (Corner 1999).

The organic aIproach is crucial for both *landscape urbanism* and organic parks, a conceptualized imagination for projecting large-scale urban parks in the future. Largely associated with *landscape urbanism*, the North American park design paradigm has an organic identity and is based on ecosystem dynamics and processes. They will be completely expounded in Chapter 4.

*3.3. Landscape Structrualism in Germany*

In the early twentieth century, the concept of *structuralism* in Europe developed in the field of structural linguistics (Deleuze 2002). This concept was introduced as an essential avant-garde movement into European architecture and urban design beginning in the 1960s on account of criticizing modern functionalism. The movement of architectural *structuralism* and its influence on German *landscape structuralism* are dissected in the section on German large-scale urban parks with the *structrualistic approach* in Chapter 4. This is in that the understanding of German *landscape structuralism* is reflected in one of the essential manifestations: the structuralistic park model.

Since the 1960s, an understanding of meaningful *structure* has functioned in the field of architecture and urban design, where proponents of *structuralism* claimed:

"We are faced with the necessity of evolving structures and forms, which can develop in time, which can remain a unity and maintain the coherence of the components at all stages of their growth. The absence of this must lead to selfdestruction." (Lüchinger 1981)

Realizing the significance of *structure*, architectural theorist Arnulf Lüchinger defined the concept of *structure* as a whole of relations in which elements could shift while still remaining independent of the whole and maintaining their meaning. The elements' interrelations are more crucial than themselves; the elements are replaceable, rather than their relations (ibid.). Emphasizing the relations instead of every single element, the *structure* may offer an open system for adaptable spaces to further urban development and flexible transformation, compared with pre-establishing urban spaces for mere satisfying functions, according to modern functionalism.

Under these influences, the theoretical application of *structuralism* emerged in German landscape architecture at the beginning of the 1980s, contributing to the development of parks characterized by diverse legibility, flexible availability, and site-specific and historic links (Weilacher 2014). German landscape architect Peter Latz generated a profound impact on German large-scale landscape architecture and parks based on his predecessors' ideas of *structuralism*. According to Udo Weilacher's statement, "Peter Latz found his way to structuralism via the writings of architects like Aldo van Eyck and Herman Hertzberger, the philosopher Claude Levi-Strauss, the astrophysicist Fritz Zwicky, and the designer Horst Rittel" (Weilacher 2008). In his unique way to understand *structuralism*, the Peter Latz school of thought is defined as an urban landscape *school of thought* in the above discussion, in parallel with the James Corner school of thought.

Peter Latz's explanations of structuralism should be grasped in a broad way, from the sociocultural perspective. In this sense, structuralism is "a theoretical paradigm emphasizing that elements of culture must be understood in terms of their relationship to a larger, overarching system or structure" (Blackburn 2008). This suggests that there is always a group of key structural systems that depend on their significance behind a one-of-a-kind sociocultural condition.

Consequently, Peter Latz transferred the idea of architectural structuralism into an analytical and designing method by inventing "informational layers" (Latz 2008a). The concept of information from Peter Latz's perspective will be explained in Chapter 4. A large amount of information, including "existing, visible landscape elements" or "invisible layers of information" (which, for example, may consist of the memory of a place or be based on experience) (ibid.) is naturally rooted within the specific, complex socio-cultural context. They not only constitute the understanding of site but also "make a significant contribution to the construction of landscape" (ibid.). The deep understanding of site sociocultural history and characteristics, through the information system, may be impacted by the concept of "palimpsest" proposed by André Corboz in 1983. He outlined that "how the land, so heavily charged with traces and with past readings, seems very similar to a palimpsest" (Corboz 1983). In site transformation, a series of crucial information is regarded as "vestiges" of site that could be used as "elements, as reference points, as accents, as stimulants for our own planning" (ibid.). The conversion process is "a more intelligent intervention" (ibid.).

Moreover, the *structuralistic approach* guides the concept of landscape into a way of ongoing analysis. Peter Latz explained in a 2006 unpublished lecture, quoted in Udo Weilacher's 2008 book *Syntax of Landscape*, that "the exciting thing about this method is that the analysis becomes an integral part of the model and is not separated from the design process, as tends to be the case in landscape planning, for example". This approach has been perceived as vital in the planning and design processes owing to its advantage in holistic analysis. It has been pointed out by Udo Weilacher that the *structuralistic approach* becomes valid in large projects where the size of the site alone makes it impossible to design each square meter individually (Weilacher 2008).

In landscape analysis, the *structuralistic approach* is offered for disassembling different and overlapping analytical structural levels when a complex, built landscape system at a large scale is considered. German landscape is hereby considered Gefüge, namely, a spatial structure composed of superimposed structural levels (Weilacher 2014). Levels of water systems, transportation systems, open space systems, building structures, and additional relevant networks are separately contemplated for analytical purposes and analyzed for specific problems (ibid.).

In this chapter, the two ranges of contemporary urban landscapes, two analytical hypotheses, and two theoretical frameworks developed over time and supported by critical rationalism approaches were discussed. They are inseparable from the urban dissolution crisis caused by post-industrialization in North America and Europe. In both regions, concrete urban landscapes rely on their underlying, implicit, but strong regional cultural embedding. Such an approach is commendable concerning continuous urban landscape improvement.

In the research hypothesis, these existing debates and discussions on urban landscapes in two cultural conditions will drive their large-scale park models in remarkably different ways, leading to the latter two chapters. Similarly, two park models are bound up with social uses, ecological functions, and their own cultural identities, which are expressed in five park qualities. With the critical approach, the North American organic model of large parks within the post-industrial perspective is considered a 'large-scale infrastructural landscape' for contemporary practices of landscape urbanism. On the contrary, a mirror of *landscape structuralism* practices is Germany's 'structuralistic' model of large parks within 'large thinking' for the whole region. This model is planned and implemented with the changes in socioeconomic structures and ecological understanding.

The term "large parks" was coined in the North American academe and takes the lead in uncovering the exploration of large-scale landscape architectural concepts for urban landscapes, particularly concerning certain groundbreaking ideas on urbanism and ecology. Thus, North American organic parks are first explained in the next chapter, followed by the other two park design paradigms in Germany and China. The theoretical analysis and project statement of North American large parks are part of the first step of discussion in preparation for the comparison of North America and Germany.

In the wake of ubiquitous urbanization and globalization, a generic urbanism also inevitably emerged in China. Most Chinese cities, as the world's largest manufacturing base, are gradually deprived of their unique urban cultures and cultural spirit. Increasingly similar, featureless city images emerged. Chinese architect Ma Yansong stated that a host of "soulless shelf cities" appeared in contemporary China (Ma 2013b). The crisis of cities' cultural identity may be ascribed to the blind pursuit of profit maximization and utilitarianism, namely, the pursuit of material civilization. It reveals that the impetus of economic development has far exceeded other factors in the social transition and urban growth.

As indicated by American scholar Christopher Marcinkoski, the urbanization process that has taken place in China over the past three decades can be summarized as four characteristics: "Scale", "Speed", "Spectacle", and "Superlatives" (Marcinkoski 2020). For some Chinese scholars, his understanding and presentation of urban construction in China are possibly poignant but are also straight to the point. Undeniably, urbanization based on massive consumption has brought the development of contemporary urban landscape in China to a standstill. The urbanization process tends to emphasize an idealized yet increasingly placeless modernization by getting rid of cultural texture and ignoring physiographic, biotic, and edaphic conditions (ibid.).

In Chinese academic circles, Liangyong Wu, a leading expert in architecture and urban planning, pointed out the problems of urban-related research as early as the 1980s. He argued that despite plenty of scholars exploring traditional Chinese cities and architecture more from the perspective of building elements, little attention has been paid to the planning of their cities and the design of architectural groups. Actually, in this respect, it is starkly different from various Western theories ranging from urban theory to design methods and even philosophical thinking, which is worth continued exploration (Wu 2011). Similarly, Chinese landscape architects attempt to take a broad global—local sense of the interpretation of contemporary urban landscape and its integrated systems.

With respect to the contemporary urban landscape in China, there is a unique path to its development that is aligned with the fast-changing urban conditions and innate cultural essence of shan-shui. The following three aspects of the shan-shui city, city of beyond beauty, and sponge city may indicate the conceptual directions of development for landscape at the urban level. Whether in theory or practice, the increasingly critical landscape theories are expected to improve the understanding of landscape. This prompts our contemplation on how to envision the *dynamic landscape* in an international trend and under a regional cultural context.

### 3.4.1. Shan-Shui City: An Ideal Settlement in Complex Realities

In order to re-discover the cultural value and spatial quality of contemporary urban landscapes, Chinese professionals began to explore the nature of the city. Until the end of the last century, cities were not supposed to be living machines since "even the most powerful technology and tools can never endow the city with a soul" (Qian 1996). In 1990, the idea of shan-shui city was re-proposed by Chinese scientist

Xuesen Qian for the theoretical conception of contemporary Chinese cities based on the traditional and ideal shan-shui culture and spirit. Since 2000, Liangyong Wu has suggested that the shan-shui city would become an essential planning concept for managing a harmonious relationship between the natural environment and human settlements. In his book *An Introduction to Sciences of Human Settlements*, he explained the idea of the shan-shui city as blending the artificial into the natural mountain–river pattern (Wu 2001).

In the traditional shan-shui culture, the concept of the shan-shui city, dating back to ancient times, is an expression of mountain–water worship. It followed the Chinese ancient politician Zixu Wu's spatial strategy of locating cities by observing the earth and examining water bodies for defense, during the fifth century BCE. Generally, the city with an ideal location is embedded in a natural shan-shui context. The shan-shui city, a long-standing urban model, is heavily influenced by the traditional theory of "Feng-shui", traditional Chinese geomancy, literally "wind–water" ("风水" in Chinese), recognized at the beginning of the Han Dynasty in 206 BCE (Shannon 2012). Ancient cities, villages, and residents surrounded by natural mountains were arranged in a desirable and ideal location and configuration, facing waters, and warm, south winds, according to the laws of Feng-shui theory.

In the contemporary urban context, Chinese professionals attempted to examine their lost shan-shui cultural spirit and the holistic approach. Liangyong Wu stated that "the tight integration of 'architecture—landscape—city' is the core of the traditional Chinese city design theory and methodology" (Wu 2000). However, ancient urban planners' holistic view is not fully inherited by contemporary Chinese urban planners and landscape architects. The increasingly apparent separation of regional planning and landscape planning and design allows the urban landscape to be understood and analyzed at a relatively small scale.

However, it remains necessary for Chinese professionals to carry out the systematic analyses and formulations of landscape at the urban level. Meanwhile, it is noteworthy that the discipline of landscape architecture is always claimed to supposedly take on the mission of the profession in the West, as clarified by the reflections on the contemporary urban landscapes in North America and Germany. If so, "how will the discipline and profession now respond? How should the discipline and profession respond?" (Weller and Hands 2020). These issues are also worth considering for Chinese landscape architects.

First and foremost, the urban landscape conception within the shan-shui structure represents one of the most significant paths to development for contemporary landscapes in the context of cultural inheritance. However, the shan-shui city as the ideal urban model essentially aims to shape a futuristic utopian urban landscape to be considered in critical thinking. Several professionals have become averse to adding more passion by pouring thoughts about the future into the molds of more-or-less comprehensive utopias. Pursuing an illusory landscape image may not only lead to the abandonment of the concern of urban reality but also the formation of a fixed thinking pattern. As American landscape architect Alan Berger claimed, "nothing is really wrong with ideals. It all goes wrong, though, when programmes put forth to realize an ideal are elevated to the level of dogma" (Berger 2009).

Hence, the generation of a futuristic utopian image should not be the focal point for the Chinese urban landscape, while the natural and man-made shan-shui structure is supposed to be extracted creatively from specific sites and artistic representations by urban planners and landscape architects, based on their understanding of the shan-shui city. The unique spatial shan-shui structure reveals the holistic approach to analyzing and planning sites at both regional and local scales. It embraces not only the construction of the shan-shui framework at a regional scale but also the piling of mountains and formation of waters in an artistic manner at a local scale. The natural texture as an objective reference for planning and design becomes one of the most vital landscape elements. To sum up, the construction of the shan-shui spatial structure under the guidance of shan-shui culture may turn into an essential issue for the Chinese urban landscape.

### 3.4.2. City of Beyond Beauty: The Di-Jing Conception

As for the comprehension of large-scale landscapes or landscapes at the urban level, it is supposed to become the mainstream in landscape architecture as a discipline in China, because the land or territory is supposedly the primary research subject. The establishment of the object of study in a broad sense is essential for reshaping the contemporary urban landscape, providing a crucial opportunity for re-development for the profession of landscape architecture. In fact, this point has been demonstrated by the frameworks of *landscape urbanism* and *landscape structuralism* in the West.

With regard to the urban-level landscape as conceived by taking a holistic approach, there was a movement called "The Ideal of the National Landscaping and Gardening" for transforming the whole of China into an enormous park or garden in the Mao era during the 1950s (Zhao 2010). As the idea of combining "greening" with production was introduced from the former Soviet Union, this movement was aimed mainly at mobilizing the mass to plant trees for afforestation in a relatively closed social environment. According to the Chinese landscape architect Jijun Zhao, the movement triggered a preliminary exploration of parks from the perspective of Chinese garden traditions including the mountain-and-water pattern, affection for nature, and poetic quality and pictorial flavour (ibid.). Apart from that, a number of landscape planning pioneers specializing in landscape architecture have proposed preliminary analyses of modern landscapes. For example, Zhi Chen's national "Fengjing" ("风景"in Chinese) suggested the Chinese worldview of scenery for the sake of nature conservation and inspiration, Juyuan Wang's "Land Scenery Planning", Xiaoxiang Sun's "Earthscape Planning", and Liangyong Wu's "Earthscape Theory" during the twentieth century (Yang 2020).

However, their thoughts have yet to be effectively advanced in the twenty-first century by most landscape architects. Actually, with the building of "Gardening Making Group" ("造园组" in Chinese) in 1951, there were attempts made by professionals to develop the Chinese theory of modern landscape architecture. However, there is still no clear and systematic understanding gained in China as to the modern concept of landscape architecture, as formally introduced from the West after the reform and opening up of China in the late 1970s. It is even more difficult to develop a large-scale landscape analysis and perspectives.

Based on the modernity of landscape architecture and landscape understanding expanded to a broader sense, Chinese landscape architect Rui Yang put forward the mega-concept of "Di-jing" ("地境" in Chinese) through the combination of "natural environment", "artistic environment", and "built environment". This is purposed to rethink and rebuild the contemporary urban landscapes in the twenty-first century (Yang 2020). According to him, the "Di" denotes land directly. The complex formula of "Di-jing" embodies "national land and territory", along with the various attributes in terms of natural ecology, human activities, constructed facilities, and aesthetic values (ibid.). Moreover, he indicated:

> "At the time when national spatial planning for the land has not yet been determined, landscape architecture has the potential and opportunity to clarify a research and practice agenda that brings natural ecology, human activities, artificial infrastructures, and aesthetic values together. In short, this is di (land). Showing this attitude and using it as a framework to build its own academic conceptual system and method, landscape architecture may become the core discipline responsible for the creation of ecological civilization and Beautiful China. Otherwise, landscape architecture will not only be marginalized in the future, but its survival and development as a discipline will become a serious problem." (ibid.)

In summary, in a circumstance where the exploration of urban landscape is quite limited, the "Di-jing" analysis is expected to contribute to the diversity in understanding contemporary landscape through critical thinking. Given the enormous dilemmas facing Chinese landscape architecture in the discipline, it is also crucial to develop critical theories and conceptual approaches to landscape. Meanwhile, based on a critique of city beautification as the only philosophy of design, it is necessary for Chinese landscape architects to understand that our task is to create high-quality urban spaces for urban residents, rather than just producing the idealized images of cities and pure landscapes. These relative contents have been demonstrated in the reflections on contemporary urban landscapes in North America and Germany.

### 3.4.3. Sponge City: Ecological Infrastructure and Structural Envision

Another conception of the urban landscape may point to the sponge city, the purpose of which is to introduce ecosystem services and mega-structures for the establishment of integrated blue-green infrastructure and urban spatial organization. In terms of ecological functionality, the pace of recent urbanization in China has largely precluded the practices of development that take into account ecological concerns, as argued by Richard Weller. However, the Chinese authorities decided to reverse this trend, proposing not only the slogan of "The Chinese Dream" to preserve the beautiful earth but also the concept of Park City for nature to play its role under the context of moderate human intervention. Meanwhile, as the landscape architecture profession advances in China, "environmental awareness is growing and the connections between landscape ecology and human health in relation to the design and construction of urban form are increasingly appreciated" (Weller 2020). As the core of sustainable Chinese urban development, the construction of ecological civilization is included in the process.

Prompted by shortages of fresh water and urban flooding in Chinese cities, landscape architect Kongjian Yu put forward the concept of the sponge city or ecological city in 2012 to clean and store urban stormwater and build a capacity for sustainable urban development. It is implied that both Chinese authorities and relevant professionals will face the challenge of ecological and environmental conservation in the long run. In this context, a principle of ecological priority is developed for sponge city.

The concept of sponge city borrowed the function of real sponge to give a metaphor to city. A city could act as a green sponge to improve urban functions of natural storage, permeation and purification. Using the landscape as a sponge is a good alternative solution for urban stormwater management (Yu 2012). The concept of the sponge city generally highlights the resilience of cities from an ecological perspective and signifies an increase in the ability for nature to respond to change.

The proposition of the sponge city stimulated the Chinese urban landscape from a landscape-ecological perspective. Specifically, the urban landscape as green ecological infrastructure borrowed the 1990s understanding in certain developed countries, identified as a widely recognized planning tool for natural conservation and regional and urban development.

Substantially, the comprehension of ecological infrastructure is not within the traditional scope of landscape architecture because the urban landscape has been deficient of ecological theories. A primary reason may account for that. The dominant Feng-shui theory in the traditional landscape meaning was regarded as the scientific, reasonable principle to manage the relationships between nature and humans, rather than introducing and developing other ecological ideas. However, the ancient Feng-shui theory is impossible to be completely inherited by landscape architects and urban planners, nor to tackle increasingly severe ecological and environmental issues in contemporary urban society.

Consequently, more Chinese professionals attempt to address urban ecological problems with landscape-ecological theories and make a combination with urban landscape. Kongjian Yu's formulation of the Chinese urban landscape as green ecological infrastructure at large and regional scales compensates for this blank. He proposed certain landscape strategies to protect and strengthen ecological infrastructure:

> "Maintaining and strengthening the overall continuity of landscape patterns and processes; protecting and establishing diverse native habitats; integrating the former farmland shelterbelts into urban green systems; establishing green heritage corridors that integrate environmental protection, leisure, education, and cultural heritage preservation and that include areas along gorges, channels, roads and railways; integrating parks into cities as the basic means of achieving high-quality life." (Yu 2012)

In fact, there are many landscape architects in the West who advocate the idea of *Porous City*, giving recognition to the conception of the city as a sponge with porous green fabric. In this sense, the new grounded knowledge of "porosity" can be found in the dimension of cities from both developed and developing countries. For most landscape architects in the developed world, the concept of the sponge city is viewed more as a landscape system. This necessitates a structure capable to ensure both the flexibility and stability of living tissue. In the view of Udo Weilacher, the sponge city plays a role in connecting the centrally important theoretical approach of structuralism (Weilacher 2018), as mentioned in the analysis of the German urban landscape—*landscape structuralism*. Due to the methodical application of structuralism, trail-blazing open space projects have been developed, which feature extensive readability, flexible utility, site-specific historical references, and high social permeability (ibid.).

Therefore, with regard to the sponge city, it can be viewed by Chinese landscape architects from the perspective of integrated ecological infrastructure, for flexible reaction to the changing environmental influences. Additionally, with this notion as the starting point, it is possible to conduct the conceptual analyses and synaxes of current Chinese urban landscapes in a structuralistic way.

To sum up, there are four key points to present as follows for the development of the Chinese urban landscape:

Firstly, the gradual transition of the overall spatial structure will be accompanied by the increasing integration of urban and rural spaces. As for the theoretical understanding and practical construction of the urban landscape, it is necessary to expand them accordingly in the future. According to the single city beautiful standard, however, it focuses on other things than the core region of a city for so-called images. In the latter part of this work, the expanded concept of the urban landscape will be taken as one of the rethinking contents. Among them, it is definitely necessary to expand the understanding of landscape into the land and territory. Moreover, for Chinese landscape architects and urban planners, it is essential to explore and perceive plain, ordinary, and artless beauty in the wider rural space, rather than ornamental, grand, and high-maintenance beautification. The urban landscape is expected to incorporate more urban and rural elements as conceptual resources.

Secondly, despite the early recognition given by our predecessors to the deep philosophical and spiritual traditions that regard humanity as part of—instead of apart from—nature, it remains necessary for Chinese landscape architects to clarify the existing spatial approach of the shan-shui structure as applied to cities and large-scale urban parks on post-industrial sites. It exerts a strong influence on the way we envision our cities and parks in the twenty-first century. Importantly, although the shan-shui city is seen as the utopian Chinese ideal, the shan-shui structure remains the core of the Chinese landscape, which is considered to be a distinct analytical and design approach to the urban landscape. For Chinese landscape architects, they should learn how to deal with the relationships between the spatial organization, artistic representation of landscape ideal, and complex environmental issues or various on-site challenges. It is especially true when the power of the landscape is increasingly recognized by us. In the future, it will undoubtedly be crucial to better explain and apply the shan-shui structure in landscape architecture. However, a major problem remains, that is, how to transform the shan-shui structure into a concrete vocabulary and methodology of planning and design.

Thirdly, the viewpoint of the global city should take into account the "landscape" as a key element and medium of urban development. In other words, the authorities and related professions of planning and design should give recognition to the essential role that landscape plays at the urban level with various cultural features. After realizing the role of "landscape", China has proposed the idea of "Beautiful China" which is aligned with the understanding that "lucid waters and lush mountains are invaluable assets," according to The State Council Information Office. Water and mountains appear to be consistent in the Chinese cultural view of the shan-shui landscape. What concerns Chinese landscape architects is "in what ways can we imagine and give substance to the relationship between a 'national dream' and the development of our discipline?", as argued by Chinese scholar Stanislaus Fung (Fung and Wu 2020).

In recent years of inventory urban renewal, several essential urban redevelopment projects have been launched, such as Yangpu Riverfront Public Space project (Figure 27) and Shougang Industrial Heritage Park project, with the aim to demonstrate the potential and role of landscape in such metropolitan cities as Beijing and Shanghai. In these large-scale urban projects, "landscape" has played the most important role, especially when it comes to the conversion of abandoned industrial sites. For the discipline of Chinese landscape architecture, this trend might promote the development of urban landscape theories and approaches to some extent in a critical and professional way.

**Figure 27.** Shanghai Yangpu Riverfront Public Space Project, Zhangming, 2017. Source: Photo by author.

Fourthly, if we look beyond the context of traditional Chinese landscape culture, what can we learn from the international urban landscape conception? A broader urban open space structure incorporated by the idea of *Porous City* may provide us with an alternative to the observation of urban landscapes. Apparently, its essence lies in the structure. Meanwhile, to address the ongoing climate crisis and the more complex urban environmental issues, it remains necessity of sorts to enhance the urban ecological resilience of modern Chinese cities from the perspective of functionalities, which is premised on the model of sponge city. In other words, the understanding of urban landscape requires the effort made to break away from the traditional boundary and critically embrace the ecological and structural ideas of cities that arise from developed countries with a global vision. The constant development of cities with comprehensive and resilient structures will demonstrate its intrinsic value through the restoration of landscape and the public amenity of a functional, healthy ecosystem.

To sum up, there are various contemporary urban landscape conceptions, conceptual possibilities, and developmental directions in the context of urban complexity and dynamics. To date, the urban landscape conceptions in developed nations have evolved into complete landscape ideas and analytical methodologies that reflect a systematical and critical way of thinking. By contrast, China remains at the stage of exploration. They are closely associated with "a way of seeing" or situational proprieties, and there are various urban contexts in which cross-cultural discussion is conducted. By means of urban landscape understanding, all landscape architects across the world can explore large-scale urban parks in different ways. In Chapter 4, three design paradigms are introduced regarding contemporary expansive urban parks, covering the design ideas, approaches and projects of park landscapes at the urban level.

# 4. Three Design Paradigms of Large-Scale Urban Parks

"What parks are, how they look, and the role they play in cities have advanced significantly since the time of Frederick Law Olmsted [ ... ]. Time have changed. Confronting ecologically and culturally disturbed sites such as landfills and deindustrialized parcels requires designers today to challenge the green veneers of the past and advance alternative strategies that ecologically heal the landscape while retelling the complex narrative of a place [ ... ]. Today, parks are not just *for* us but also *by* us." (Czerniak 2022)

## 4.1. Organic Parks in North America

The cross-cultural study of three large-scale urban park design paradigms takes into account current views of urban landscapes in North America, Germany, and China, as well as particular organizational parks and planning projects for reconstruction, redevelopment, and transformation. Such views lead to the discussion's beginning point, which is North American organic parks. Organic parks are the process-oriented park model for the integration of urban infrastructure and dynamic ecosystems, which is a metaphor for the North American urban environment as ecological insights become more and more pervasive. Currently, organic parks are used as a guide in many urban projects in North America to provide a conceptual framework between urban form, dynamic environmental processes, and daily life.

In a creative cultural setting, organic parks—a primary and distinctive model of modern large urban parks—will be examined. Regarding the study context, James Corner offered a forceful manifestation of the *cultural imagination* in 1999 with his critical method of *critical thinking* to comprehend the contemporary urban landscape of North America. With evolving concepts of urbanism and ecology, organic parks will be designed and certified in terms of size, social, and ecological features from both quantitative and qualitative viewpoints. This is because the *cultural imagination*, which contains unlimited creativity, is a cultural embedding.

### 4.1.1. Vision: Creative Cultural Comprehension

An innovative large-scale urban park paradigm with an organic approach was established by the academic community of North American landscape architecture to meet the global challenge of site transformation, particularly on primarily filthy and polluted abandoned industrial land. This section verifies how the production of a unique definition is firmly based in its own creative cultural setting.

There are roots for views on individual freedom in North American thinking. As a result of their shared commitment to individual liberty, a bond between them developed and eventually took the name "American Spirit" (Ma 2013a). In order to fully comprehend their culture, character, and conduct, it is important to value the production of original ideas. In North American society, the capacity for innovation is stimulated by a regard for freedom and independence of opinion.

From a creative perspective, strong beliefs exist in North American professionals with respect to progress in interpretations of landscapes (Höfer 2013). Specifically, these professionals are scholars who lay a solid foundation for urban landscape development, such as J. B. Jackson, Ian McHarg, Denis Cosgrove; proponents of *landscape urbanism*, such as Charles Waldheim, James Corner; and initiative promoters of *large park* model, such as Julia Czerniak, George Hargreaves. Generally, these professionals nurtured a creative theoretical environment, and their views were related to the concept of large-scale urban park on post-industrial sites. Notwithstanding, the creative cultural setting for the emergence of large-scale urban parks is considered in James Corner's viewpoint of *cultural imagination*.

In James Corner's *critical thinking*, contemporary North American landscape is "first a cultural construct, a product of the imagination" (Corner and Hirsch 2014). The landscape is recovered as "a critical cultural practice" that "enriches cultural world through creative effort and imagination" (Corner 1999). He believed that an emphasis is shifting from "landscape as a product of culture" to "landscape as innovative cultural agent" (ibid.). In other words, the landscape itself is "not simply a reflection of culture but more an active instrument in shaping and enriching contemporary culture" (ibid.).

The "creative effort and imagination" (ibid.) argued by James Corner are intimately linked to J. B. Jackson's innovative analyses of "dynamic system of manmade spaces" in 1984, and postmodern re-interpretation of space by British cultural geographers Denis Cosgrove and Stephen Daniels in 1988. Concerning the former, J. B. Jackson's contribution to the evolving understanding of urban landscape has been remarked on in the previous chapter. Regarding the latter, James Corner in 1999 quoted the following paragraph in *Iconography and Landscape* to elaborate on the nature of landscape.

> "From a postmodern perspective, landscape seems less like a palimpsest whose 'real' or 'authentic' meanings can somehow be recovered with the correct techniques, theories or ideologies, than a flickering text displayed on the word-processor screen whose meaning can be created, extended, altered, elaborated and finally obliterated by the merest touch of a button." (Cosgrove and Daniels 1988)

James Corner advocates a postmodern idea of space. Within this framework, the genuine development of North American urban landscapes more effectively created, changed, and compounded spaces than the inherently physical characteristic to obtain site information, organize specific space, and form structural connections between spaces based on multiple landscape ideas. The two different facets constitute the main conflict between the understandings of North American and German landscape, which are distinguished in Chapter 5.

The postmodern perspective supports James Corner's essential landscape understanding that it is an integration of "idea and artifact" (Corner 1999). The landscape is constructed by both imagined and material parts. For instance, Pyxbee Park (12 hectares), which George Hargreaves in collaboration with the artists Peter

Richards and Micheal Oppenheimer created on a Palo Alto landfill site on the edge of San Francisco Bay, serves as a metaphor for the union of man and nature in an artistic and organic way. In terms of ecological clean-up and restoration of the site, Hargreaves starts by covering the site with a general sandy soil and clay to prevent the exposure of waste and to create a suitable environment for weeds to grow. In order to prevent the growth of woody plants and their roots from destroying the buried soil and causing the release of harmful substances, no large trees were planted on the site, only native herbs and naturally growing weeds were allowed to grow. As seen in Figure 28, in the integration of art and materials, Hargreaves envisions a creative and aesthetic relationship between nature and the original abandoned site through carefully shaping the rhythmic topography providing the function of controlling the wind direction and sunlight in the idea of wilderness aesthetic. In addition, this park expressing sculpted earth is combined topography, vegetation, site history, and art installations and fosters a distinctive post-industrial landscape. It also reflects the understanding of landscape with the *cultural imagination* of James Corner:

**Figure 28.** Pyxbee Park in Palo Alto, Hargreaves Associates, 1991. Source: Photo by ©Shuang Zhao 2018; used with permission.

"Only through a synthetic and imaginative reordering of categories in the built environment might we escape our present predicament in the cul-de-sac of post-industrial modernity, and the bureaucratic and uninspired failings of the planning profession." (Corner 1999)

Denis Cosgrove's studies from the 1980s can be used to explain another source of *cultural imagination* in addition to the postmodern reinterpretation of space. "Landscape is thus a way of seeing, a composition and structuring of the world," he claims (Cosgrove 1985). His understanding made it possible to incorporate the unique, imaginative, and creative human experience into studies of the geographical environment. James Corner's interpretation of how we perceive the modern landscape is "eidetic and subjective" (Corner 1999), and is based on "a way of seeing".

Since the 1990s, the field of North American landscape architecture has been exposed as having a basic landscape idea of picture-making that disconnects heterogeneous representations. The imaginative activity of many landscape architects was strongly backed in order to reverse the stereotyped mechanism between the idea and representation in the shape of the charming and rural environment. Against this background, James Corner reiterated "how one 'images' the world literally conditions how reality is both conceptualized and shaped", and he believed that landscape is simply "as a scenic object, a subjugated resource, or a scientistic ecosystem" without *cultural imagination* (ibid.).

For instance, Peter Walker, an American landscape architect, created the landscape of the 9/11 Memorial with the idea of "Reflecting Absence", which offers a space for reflection and remembering. The largest man-made waterfalls were placed in two enormous voids left by the Twin Towers, reflecting the designer's understanding and imagination of public places, sites, events, monumentality, and temporal transitions. The outcome is a thought-provoking manifestation of landscape architecture. As seen in Figure 29, people experience the voids with materialization and imagination to be extremely tremendous. The relationship between "a way of seeing" and concepts of the landscape is therefore unbreakable.

**Figure 29.** The 9/11 Memorial in New York, Peter Walker and Micheal Arad, 2004. Source: Photo by ©Zhenkun Gan 2019; used with permission.

Following the idea of landscape as "a way of seeing", individual creativity embedded in ideas has been remarkably affirmed as a result of a shifting attitude toward knowledge. According to American architectural theorist Michael Speaks, knowledge is no longer concerned with absolute truth, nor does it follow a fixed and changeless idea. Rather, it is concerned with "plausible truths". Specifically, it is "no longer dictated by ideas or ideologies nor dependent on whether something is really true, everything now depends on credible intelligence, on whether something might be true" (Speaks 2006). Knowledge becomes "design intelligence" (ibid.) in the North American academe. This transformation would stimulate the generation of more creative ideas.

With the above cultural background, contemporary North American landscape can shape and enrich contemporary culture when it is considered "imaginative and material practice" (or "a way of seeing and acting") (Corner 1999). Contemporary cultural ideas in landscape architecture are potentially driven through dynamic interactions between imagination and materialization. Regarding organic parks, "changing ideas of nature, wilderness and landscape continue to inform the physical practices of designing and building in turn, further transform and enrich cultural ideas" (Speaks 2006).

In conclusion, the concept of organic parks is conceived by expounding new perspectives, composed of two newly emerging ideas on urbanism and ecology. Furthermore, the new perspectives with a critical attitude towards pastoral landscapes contain four concrete aspects, contrary to stereotypical perspectives of conventional nineteenth-century parks.

### 4.1.2. Conception: Two Newly Emerging Ideas

The large-scale urban parks in this book are further interpreted as organic parks, influenced by the term *"large parks"* used in North American academic circles, in order to highlight the interaction of the fields of landscape architecture and urban ecology based on an inherent understanding of contemporary dynamic and organic nature.

The term *"large parks"* is derived from a series of analyses and debates on contemporary large-scale parks. They include the 2003 conference, titled "The Large Parks: New Perspectives Conference," which was held at GSD, and subsequent GSD students' studies on specific park cases. At this conference, five essential aspects surrounding *large parks* were identified: "parks and site history: the made and the remade"; "parks and the city: the urban, the peripheral, and the post-urban"; "parks and ecology: sustainable design and maintenance"; "parks, processes, and place"; and "parks and the public" (Fulton 2003). These five aspects are regarded as the most noteworthy parts from which certain key points for organic parks are deduced in the following discourse.

The first facet of the made and the remade suggests that large parks are "not simply natural or found places; they are constructed, built, and cultivated—designed" (Corner 2007). Connected to the postmodern perspective, organic parks are certainly shaped by integrative forces of both human and nature. The other four facets may be categorized into two crucial points: urbanism and ecology in a renewed sense. They are considered two newly emerging ideas defining organic parks "in flux" collectively (Czerniak 2001).

Generally, the recognized landscape-based urbanism, that is, urbanism shifting towards landscape, touches upon landscape urbanism advocated by Charles Waldheim in *The Landscape Urbanism Reader* (Waldheim 2006). In Meg Studer's interview published in 2012, Charles Waldheim expressed today's *landscape urbanism* as "the question of energy, resource extraction, production, and flows in relationship to urbanism" (Studer 2011). "Landscape urbanism aspires to build an understanding of urbanism in which ecological forces and flows supporting urbanism are considered part of the city as opposed to external to it" (ibid.). Therefore, he considered the renewed urbanism a response to criticizing older models of urbanism in which a city is distinct from the countryside, and viewed energy and substance as externalities to city problems, which made a city vulnerable.

Aside from Charles Waldheim's argument on the understanding of urbanism, James Corner offered another explicit statement. Given the nature of dynamic and process-oriented urbanization happening in current North American cities identified by urban planners and landscape architects, he conceptualized "a more organic, fluid urbanism" (Corner 2006) as horizontal urban surface strategies. His further assumption of urbanism plays an immediate part in the production of the concept of organic parks. The feature of "organic, fluid" is connected to the organic approach.

In their interpretations, the updated urbanism concept is an urban landscape phenomenon with the characteristic of horizontal urban sprawl, a reasonable expression proposed by Alan Berger in 2006. Through Charles Waldheim's aspiration of landscape as a "medium" of a city, the urban landscape would remove conventional boundaries between city and nature, and city and countryside. Rather, the urban landscape is bound up with fluid and continuous urban "surface" or "field", on which the "complex interweaving of natural ecologies with the social, cultural, and infrastructural layers of the contemporary city" is established conceptually (Waldheim 2006). Organic parks cross the spatial line and will stretch out on this wide urban surface to display "landscape as urbanism" fully (ibid.). Projects of Downsview Park and Freshkills Park are representative of the trend, offering "the most fully formed examples of *landscape urbanism* practices to date applied to the detritus of the industrial city" (ibid.).

Another newly emerging idea of ecology is provided in Ian McHarg's *Design with Nature* in 1969. Undoubtedly, it has maintained a profound influence. Since its publication, "landscape architects have been particularly busy developing a range of ecological techniques for the planning and design of sites" (Corner 2006). For instance, state-of-the-art ecological restoration and engineering techniques were adopted to construct the Freshkills Park. Associated with ecological techniques, "Sophisticated Engineering Systems," including "Leachate Management System," "Landfill Gas Collection System," and "Capping System," are designed to collect and treat leachate, methane, and byproducts of waste decomposition, as well as to insulate all contamination (NYCDPR 2021).

Ian McHarg's historical importance is acknowledged by proponents of *landscape urbanism* for promoting the development of ecological techniques. In particular, he brought landscape architecture into "broader visibility as a productive practice essential to 'solving' environmental 'problems'" using a "deterministic approach to ecological and land-use planning" (Corner and Hirsch 2014). Under this premise, landscape urbanists further broaden Ian McHarg's approach to the role of *cultural imagination* in landscape architecture. They declared a conceptual approach of the "matrix" as a dynamic framework. Simultaneously, they developed an understanding of "space-time ecology that treats all forces and agents working in the urban field and considers them as continuous networks of inter-relationships" (Corner 2006). The emerging ecology embraces energy, substance, and their interactions pertaining to the living and organic urban surface, and is increasingly regarded as central in reconceiving city and urban landscape.

The ecology reflected in *landscape urbanism* illustrates that based on Ian McHarg's concept, the new ecological idea has emerged since the 1980s and was introduced into landscape architecture and organic parks in the 1990s. The development of the ecological viewpoint presents a paradigm shift from *equilibrium* to *non-equilibrium* or dynamic flux in ecosystems, with "climax community" being questioned in the 1950s (Pollak 2007).

In the *equilibrium* paradigm, the concept of "climax community" was employed by American ecologist Frederic Clements in his 1916 work *Succession*. It indicates that an ecological community may finally reach a steady state through a process of ecological succession. The "community" maintains the *equilibrium* condition until a disturbance happens. In this conceptual situation, the key point is the disturbance, which is considered external to the ecosystem (Rosenberg 2007). Frederic Clements's concept of "climax community" dominated ecological research for the first half of the twentieth century before it was questioned (ibid.).

In the 1980s, the *equilibrium* paradigm was challenged "by statistical and probabilistic approach that have revealed disturbance to be a frequent, intrinsic characteristic of ecosystems" (ibid.). In other words, considerable scientific studies have been conducted by relevant scholars to redefine the model of ecosystem development in the face of an untenable assumption of ecosystem, excluding the *disturbance*. Among them, ecologist C. S. Holling's dynamic model was proposed in 1992 to reinterpret the nature of ecosystems. This will be further explained in this chapter. In this model, *resilience* pertaining to ecosystems become an inherent property showing its adaptive capacity. The notion of *resilience* is associated with the urban landscape's new sustainability. It is among the characteristics of organic parks, added to the qualitative analysis.

In short, the *non-equilibrium* paradigm "reframes nature in terms of its continual disturbance, rejecting the previous scientific 'truth' of organic nature's tendency towards either equilibrium or homogeneity" with the scientific research and published evidence in the field of ecology (Pollak 2007). The *disturbance* has been accepted as a part of ecosystems in the *non-equilibrium* paradigm:

> "We've seen the paradigm of ecology move toward a more organic model
> of open-endedness, flexibility, resilience, and adaptation and away from
> a mechanistic model of stability and control. In other words, ecosystems
> are now understood to be open systems that behave in ways that are self-
> organizing and that are to some extent unpredictable." (Lister 2016)

Two newly emerging ideas of urbanism and ecology that thrust the concept of organic parks are associated with the *landscape urbanism* framework. Regarding the two emerging ideas, the renewed ecological consideration of *non-equilibrium* paradigm certainly offers essential clues, contributing to the remarkable development of both *landscape urbanism* and large-scale urban parks toward an organic approach. However, these prominent changes in urbanism and ecology are also intertwined with systems theory, leading to a paradigm shift in our understanding of the complexity and dynamism of the urban fabric (Capra 1996). In conclusion, the understanding of organic parks is largely expanded by incorporating urbanism and ecology with contemporary urban landscape.

Aside from these emerging ideas, new perspectives on organic parks are also analyzed by comparing them with stereotypical perspectives of conventional parks in North America. The contrast between these two perspectives elicits the following contents.

### 4.1.3. Transformation: From Stereotypical to New Perspectives

According to the historical understanding of large-scale urban parks in North American cities, the conventional park refers to the public park model that has been employed since the nineteenth century, inspired by the eighteenth-to-nineteenth-century English landscape garden (Jackson 1984). The concept of conventional parks is influenced by J. B. Jackson's "Landscape Two" and is understood as a "static" (ibid.) and "scenic object" for a subject (Corner 1999). It is positioned as an "end-product" (Marton 2010) of an organized ideal in the industrial society, where cultivated order made the city beautiful (De Jong 2000).

The conventional park is acknowledged universally as classical pastoral landscape, while it is inhabited by city dwellers of various social backgrounds within the liberal democratic and urban society (Czechowski et al. 2015). Parks realized a combination of beautiful, harmonious scenery, urban, and social functions during the industrial stage. Since they provided healthy green spaces for people's recreation and activities, American sociologist and designer Galen Cranz defined them as a "Recreational Facility" after the 1930s. These parks also stopped the spread of disease, reduced class conflict for social equity, integrated immigrants, and even educated people (Cranz 1982). Parks represented a healthy environment, a recreational facility, an experience of urban nature, and maintenance of democracy and civilization. Briefly, the conventional park is the pastoral landscape integrating scenery with functionality.

As urban green space, the conventional park played a role in offering "relief" from industrial cities (ibid.) due to the binary opposition or separation between nature and built-up urban areas. In the understanding of the static image of green space, the conventional park without doubt presented an ideal and visual harmony and emphasized its pastoral pictorial sense. From the aesthetic perspective, conventional parks are formed as a way of pictorializing nature, demonstrating an entrenched mechanism of two-dimensional landscape representation that facilitates the conventional park design to be a representative design.

However, the representative design is increasingly expected to be adjusted owing to its two possible disadvantages. First, the park design would only reproduce outdated scenic images with no reference to the changing urban and social contexts (Höfer and Trepl 2010). It forces the conventional parks to constantly be in an ideal state. Second, designing the classical harmonious landscape seems to present parks within a stable context. Once the parks are formed, transforming them to adapt to any other unaccounted factors in design processes is slightly difficult. This situation is also explained as lacking natural disturbance in recent ecological findings and analyses (Pollak 2007). The sudden *disturbance* is a distinguishing feature of living ecosystems as they are "evolving discontinuously and intermittently" (Lister 2007) and regenerating the ability of self-organization to adapt to the sudden situation. This natural characteristic is introduced into the concept of large-scale urban parks. This characteristic is distinguished as one of the qualities of organic parks and is referred to as resilience. In this sense, the organic park design is an "adaptive ecological design" (ibid.).

However, the development of fresh perspectives for organic parks is accelerated by the modern complex urban context. According to Martin Prominski, traditional parks and parks of the twenty-first century are essentially two different and opposing ideas:

"On one side, we find the conventional ideas of a harmonious, green landscape opposing built-up areas; on the other side, there are new ideas that avoid any oppositions and try to integrate the strange mixtures of our peripheries or the web of infrastructure lines which are the landscape of our contemporary culture." (Prominski 2010)

Despite their co-existence, an integrative idea is substituting for an opposite one in park conceptions. As a result, organic parks are conceived as "complex systems" in space and time (Lister 2007). This is a concept learned from ecologists based on their explanations as regards complex living ecosystems in nature. The systems are an open-ended network of structure (elements), dynamic processes, and their relations (ibid.). They simultaneously integrate infrastructure, "new" ecology, and life. They could "organize objects, spaces, and the dynamic processes and events which act upon" (Corner 1999) connected and open-ended urban surfaces.

In this regard, the concept of organic parks assumes the role of "complex medium", which is capable of articulating relations between urban infrastructure, public events, and indeterminate urban future for large post-industrial areas (Waldheim 2006). Organic parks generally bear these interlacing relations to implement the ongoing transformation of sites, rather than playing a part in the opposite of built-up areas.

Moreover, the positioning of parks shifts from "end-product" in the industrial society to "work-in-progress" in the post-industrial society. The comprehension of organic parks in a dynamic fashion is to "develop a process-orientated definition" (Prominski 2005). The construction of organic parks increasingly requires a consideration of its sustainable ecological function in the long term. On this basis, North American professionals believe that parks in processes are more capable of improving the city's environment in a continuous way (Marton 2010). Hence, the concept of parks changes from a static scenic to dynamic process.

Specifically, urban nature and life embody not only essential elements but also processes in the conceptual spatio-temporal systems. Urban nature means accepting natural processes and natural disturbance, whereas urban life in the everyday world reveals features of freedom, diversity, and unpredictability in programmatic processes. Natural and social processes constitute the process-orientated landscape automatically.

In contrast to conventional parks' representative design, organic parks are performatively designed, design that emphasizes the "performance" of large-scale urban parks (Czerniak 2001). The term *performance* is derived from linguistics, and its meaning is always connected to certain behavior and action. The *performance* of physical material indicates shifting the focus of interest from essence to effect. Thus, the key issue is not "what things look like", but "what they do" (Allen 1999). Transferring it into organic parks, the idea profoundly uncovers the concern of ecological effectiveness or functionality.

North American professionals supported organic parks' performative design when they managed messy, derelict, and contaminated post-industrial lands. Since these lands hold certain complexities in the actual urban environment, it is unfeasible to implement, cultivate, and transform them overnight. In this situation, the "pragmatic and processual view" (Prominski 2010) is essential to organic park planning and design. This view suggests that the ecological effectiveness will show and grow over time, representing the site transformation from a pragmatic point of view.

From the new perspective, organic parks on urban surface essentially attempt to "create an environment that is not so much an object that has been 'designed' as it is an ecology of various systems and elements that initiate a diverse network of interaction" (Corner 2006). In conclusion, Table 3 presents the transformation from conventional parks' stereotypical perspectives to organic parks' new ones by means of comparison.

**Table 3.** Organic parks' new perspectives compared with conventional parks' stereotypical perspectives in terms of concept, positioning, role, and focus.

| Category | Stereotypical Perspective | New Perspective |
| --- | --- | --- |
| Concept | 1. A "static", "scenic object" (Corner 1999); 2. The pastoral landscape integrating scenery with functionality | 1. "Complex systems" in space and time: an open-ended network of structures (elements), dynamic processes, and their relations (Lister 2007); 2. The process-orientated landscape containing natural and social processes |
| Positioning | Park as "end-product" in the industrial society (Marton 2010) | Park as "work-in-progress" in the post-industrial society (Marton 2010) |
| Role | Green space for offering "relief" (Cranz 1982) from industrial cities or built-up urban areas | A "complex medium" (Waldheim 2006) to transform post-industrial sites |
| Focus | Representative design: a static and ideal image or visual harmony in a pastoral pictorial sense | Performative design: emphasizing large-park "performance" (Czerniak 2001) in the "pragmatic and processual view" (Prominski 2010) |

Source: Author's compilation based on data from Cranz 1982; Corner 1999; Lister 2007; Marton 2010; Waldheim 2006; Czerniak 2001; Prominski 2010.

## 4.1.4. Approach: Critical Thinking

In the second chapter, the comprehension of critical rationalism approaches employed in the research has been construed. The approaches in the field of landscape architecture concern a reflection, entailing a thoughtful analysis of the issues and values involved. Moreover, the North American academe has its own interpretations of the critical approach.

Since the 1980s, a reconsideration of making connections between the established theory of landscape architecture and critical approach has emerged as J. B. Jackson analyzed the term 'landscape' critically. In the early 1990s, there was continuing debate on the nature of theory in landscape architecture (Swaffield 2006). An organized discussion on North American landscape architecture and critical reasoning was performed at the conference of the Council of Educators in Landscape Architecture (CELA) in 1990. In the discussion, influential arguments for *critical thinking* originated from James Corner's *critical thinking* of "creative processes" in 1991 and American landscape architect Elizabeth K. Meyer's criticism of "either-or" in 1997.

Concerning the necessity for critical thinking in the theoretical analysis of landscape, North American scholars realized that this standpoint assumes an essential "point of view within a theoretical infrastructure" (McAvin 1991) for landscape rethinking and readjustment. Specifically, the perspective of critique plays an essential role in resisting the taken-for-granted ways of thinking and thrusting alternatives (Swaffield 2002). The alternatives articulated are triggering creativity for James Corner, but a collapsing established duality for Elizabeth K. Meyer. Consequently, both scholars impel the advancement to a more appropriate landscape understanding, laying a theoretical foundation for the organic park concept.

Creative Processes

At the 1990 conference, James Corner proposed: "What is critical inquiry? What does it mean in the context of landscape architecture?" Then, in his 1991 essay Critical Thinking and Landscape Architecture, he indicated that *critical thinking* should incline to creativity exhibited by implementing a new working pattern of *plotting* in landscape design. The procedure of *plotting* a land includes building "a piece of ground" (as "physical sites constructed"); "graphic representation" (as "eidetic sites imagined or thought" in a map or plan); constructing "a narrative or time series" (as "future sites" in an unfolding, sequential plot); and strategic devising of a plot (as "inherited sites") (Corner 1991). Landscape in a specific time and space is created critically through this pattern.

For James Corner, the *critical thinking* of landscape is combined by conceiving sites as creative processes. He stated that "we map and 'lay out' our agendas and strategies, connecting and revealing previously unforeseen relationships. To plot is to critically cultivate our relationship to landscape" (ibid.). In the creative process, the key is to construct relationships between possibilities, unpredictability in sites (urban life), form, and structure with strategies. The established relationships developing over time are considered fluid, unconstrained, and self-organized to a certain extent. The fluidity and self-organization are emphasized because social and ecological qualities in the urban landscape are manifested in self-organization in uncertain urban life and complexity in the landscape-ecological sense. Such a pattern affects James Corner's concept of organic parks in planning and design, such as Freshkills Park.

Criticizing "Either-Or"

Elizabeth K. Meyer criticized "either-or" for representing a rigid and outmoded binary thinking pattern in her 1997 essay The Expanded Field of Landscape Architecture. She questioned: "Why do landscape architects so frequently describe the world and their work in pairs of terms? Either-or. This or that. One or the other" (Meyer 1997). The production of binary thinking pattern has been essentially identified by certain philosophers and cultural critics as "a tool for controlling power and making natural hierarchical relationships" since classical times (ibid.). However, the thinking pattern is not any more suitable for the comprehension of contemporary landscape architecture. By analyzing Elizabeth Meyer's *critical thinking*, the way of thinking can be adjusted and improved to contribute to the advancement of landscape understanding. As she stated, such advancement will usher in the arrival of an expanded field of landscape architecture.

From Elizabeth Meyer's viewpoint, the significance of rejecting 'either-or' lies in "avoiding destructive polarization" (ibid.). The 'either-or' division largely destroys the interconnections and interactions of binary landscape elements, such as city and landscape; urban and rural areas; culture and nature; art and science. As a result, the understanding of landscape falls into a simplistic and fixed view. Hence, the interrelationships of binary landscape elements should be reconsidered and reconstructed in a new form. A strategy of "in between" is supported, which may induce their interrelationships to be complex and diverse. In other words, the space between the binaries should be discovered. Thus, the strategy leads to a landscape concept expanded as "hybrid, continuum or cyborg" (ibid.).

The above opposite elements represent an "exclusive differentiation" (Beck and Lau 2005), presenting an absolute difference and cutting off possibilities of ties and reciprocities simultaneously. However, a shift from an "exclusive differentiation" to an "inclusive differentiation," in which categorizing is plural and ambivalent, exists in the hybrid understanding. The hybrid becomes an alternative way of seeing and describing the North American landscape under critical thinking.

The hybrid, continuum or cyborg is probably linked to spatial ideas of postmodernism, as well as machinic assemblage within the theoretical framework of *landscape urbanism*. From a postmodern perspective, an image of intersections, overlaps, hybrids, and cyborgs is created only by acknowledging that binary terms can be related to one another without implied hierarchies or dominances (Huyssen 1986). Additionally, the key term *machinic assemblage*, which is a concept similar to Meyer's hybrid, was proposed by Mohsen Mostafavi and Ciro Najle in their 2003 book *Landscape Urbanism: A Manual for the Machinic Landscape* (Mostafavi and Najle 2003). In 2004, French philosophers Gilles Deleuze and Felix Guattari reinterpreted that *machinic assemblage* should be employed in a more free, open-ended way, in which various elements interconnect and assemble (Deleuze and Guattari 2004).

According to the analysis of Elizabeth Meyer's criticism of 'either-or', the expanded concept of landscape may produce an effect on the definition of North American organic parks. Through "the lens of size", a critical perspective is generated because North American professionals attempted to "cut across conventional binary categories of classification, historic or contemporary, built and unbuilt, competition-sponsored or commissioned" (Czerniak and Hargreaves 2007). Furthermore, they set insights on the impact and significance of size relative to the planning, design, and management of parks, past and future. With size as a new definition, the analysis of organic parks breaks down limitations of binary thinking.

### 4.1.5. Qualification: Quantative and Qualitative Analysis

The concept of organic parks suggests that size becomes an essential "premise" (Czerniak and Hargreaves 2007). Since 2003, the term "large" as a singularly important criterion for size (Lister 2007) has begun to define parks explicitly for the purpose of opposing the traditional binary thinking. Thus, the North American large-scale urban parks are primarily analyzed from a quantitative perspective.

However, "large" means more than quantity. Beginning with the size, it also takes in another two connotations, implying the role of participating in shaping urban horizontal surfaces (Wall 1999) and a multiplicity of social and natural concerns (Pollak 2007). The former is reflected in organic parks as "extensive landscapes integral to the fabric of cities and metropolitan areas" (Corner 2007), with a North American landscape ambition. The latter suggests the multiplicity of natural and social concerns in the urban landscape, as reflected in organic park heterogeneity.

The size and the two meanings stated above are bound to bring about organic park qualitative changes under social and ecological considerations. This suggests that only the quantitative perspective is not sufficient for its qualification. Thus, a qualitative perspective is also required in the following analysis. Qualitative research is a traditionally empirical approach in the field of social sciences. It helps to understand that the concept of organic parks is limited not only to an absolute quantitative criterion but also orientated toward relative qualities and values for contemporary cities, ecology, and an individuals' everyday urban life.

Size Criterion and Two Dimensions

As implied by the term "large", size matters. The GSD studies on organic parks demonstrate that their acknowledged quantitative criterion is at least 33 hectares in the area within contemporary metropolitan regions (Czerniak and Hargreaves 2007). According to this measurement, the analyzed organic park practical projects, namely, 231-hectare Downsview Park in Toronto and 891-hectare Freshkills Park in Staten Island, New York, in later sections meet the criterion completely.

The size criterion could be traced back to American landscape designer Andrew Jackson Downing's proposal when he lobbied for a larger tract of land for Central Park in the mid- to late 1800s (Czerniak and Hargreaves 2007). Owing to a fear of health issues associated with unrelieved density in the early stages of America's urbanization (Cohen 1997; Czerniak and Hargreaves 2007), he proposed:

> "Five hundred acres are the smallest area that should be reserved for the future wants of such a city [ ... ]. There would be space enough to have broad reaches of park and pleasure-grounds in that area, with a real feeling of the breadth and beauty of green fields, the perfume and freshness of nature." (Olmsted and Kimball 1928)

Organic parks as an essential type of urban green open space called for sufficient land to satisfy recreational function for a collective and perfect urban hygienic
environment, offer aesthetic perception, and improve one's mental and psychological state. Moreover, the large tracts at that time could be arranged for "the pleasure
ground," which is the first type of North American urban park by Galen Cranz in
1982 because generous space could be acquired easily and cheaply in reality (Rybczynski 1995).

Hence, the nature of large-sized parks during the early period of urbanization
was conceived as "an anti-urban ideal" (Cranz 1982) and served as "counterparts to
cold, technical modernity" (Prominski 2010). Parks represent the subjectivity of the
object (nature) and require large areas of land to organize the picturesque landscape
with an image of greenery for expressing a harmonious relationship between nature
and humans. The ideal relationship alleviated the complications of urban life in the
industrial society.

However, the quantitative criterion of parks is not static. Jane Jacobs had criticized this criterion since the early 1960s. Influenced by the urban crisis, the largeness
of parks was considered a "liability" (Czerniak and Hargreaves 2007). The urban
crisis in North American cities manifested the problems of metropolitan growth. It
triggered the flight of the middle class from failing cities to the suburbs; as a result, people no longer sought park services and even avoided parks. Such an event
prompted the realization that vast urban parks with a single use do not bring about
the attention of cities in spite of their largeness. In this situation, "dispirited border
vacuums" emerged in urban parks (Jacobs 1961). Jane Jacobs required those parks to
"make greater use of its perimeter" for stimulating residents' uses, as demonstrated
in *The Death and Life of Great American Cities*. Specifically, bringing various uses from
deep within the park to its edge may produce "spots of intense and magnetic border
activity," creating a lively connection between the park and city (ibid.).

Jane Jacobs's arguments suggest that largeness alone may be insufficient to jus-
tify the existence of parks. This criterion does not guarantee that parks have sustainable development in the city. In Jane Jacobs's opinion, a park's perimeter is as
crucial as its interiority. She treated the park as a social space in which rich urban
activities and programs for a wide range of uses could be introduced into both the
perimeter and interiority. In conclusion, the understanding of size as large is not
only placed on the review of historical factors but also new interpretations, namely,
the two connotations of size.

Two potential connotations of size embrace the role of participating in shaping
urban horizontal surfaces and a multiplicity of social and natural concerns. Generally, they constitute two kinds of analytical dimensions of organic parks: extensive
and inclusive.

The extensive implies that the concept and organization of organic parks unfold at a horizontal level of urban surfaces as the park's size increases. In this sense,
organic parks are even called "extensive landscapes" (Corner 2007). The inclusive
suggests that the park area should be big enough, in excess of 33 hectares, to contain richer resources and landscape elements at multiple scales for the park's own
making (Czerniak and Hargreaves 2007). Thus, organic parks could establish more
interconnections with an urban nature and urban life.

- Extensive Dimension

First, two questions should be raised for extensive dimension. Why are organic parks discussed in the extensive dimension? Why are they capable of shaping urban horizontal surfaces?

The answers have been primarily related to an emerging feature of contemporary North American cities since the late twentieth century. It is identified as "a radically horizontal urbanism" (Allen 2002) and further as "predominantly a horizontal landscape phenomenon" (Berger 2006). According to Alan Berger, "the type of development found in 'sprawling areas' mainly consist of horizontally oriented landscape planes and surfaces, not buildings (notably vertical density)" (ibid.). It could be summarized as thinking of landscape-based urban development in contemporary cities.

The above North American urban phenomenon during the 1980s and the 1990s is ascribed to the extension of urbanism to the whole urban areas, particularly the influence of *Post-Fordism* on the spatial organizational structure and features in North American cities. Architect Patrik Schumacher revealed that Post-Fordist production paradigms are organized around emerging principles of decentralization, horizontality, self-organization, rapid mutability, fluidity, and indeterminacy with the failure of stable cycles of reproduction and expansion (Schumacher and Rogner 2001).

Consequently, the dispersed production patterns at a socioeconomic level lead contemporary North American cities into an open, decentralized, and self-organizing spatial model. Therefore, the horizontal urban sprawl is explained rationally through the Post-Fordist mechanism. This mechanism demonstrates that "the relation between modern urbanism and Fordist economic imperatives" (ibid.) in industrial cities has been replaced by the further relation between 'landscape-based urbanism' and the Post-Fordist mechanism in post-industrial cities. For urban planners and landscape architects, horizontal urban sprawl causes "the disappearance of North American cities into landscapes" (Shane 2006).

Hence, the landscape dominates the spatial organization of urban surfaces. Meanwhile, the real root of *landscape urbanism* development begins. Considering the dominance of landscape, James Corner argued explicitly for the close relationship between urban surface and today's landscape:

> "The contribution of landscape for the twenty-first century would be providing a more primary foundation for the city—the very bedrock, matrix, and framework upon which a city can thrive, sustainably with nature and equitably with diverse cultures and programs." (Corner and Hirsch 2014)

The arguments indicate that organic parks as a concept for urban landscape planning will undoubtedly play an essential role. As a response to the emerging urban features, organic parks will "retain a large-scale sense of landscape, horizon and extension" (Corner 2009). In the planning and design, organic parks aim to shape the urban surfaces and establish "a seamless network of inter-connectivity" with the strategy of "unification" for overcoming segmentation of a given site (ibid.).

Second, the extensive dimension of organic parks could be derived in part from the ambition of urban planning at the turn of the twentieth century. This reasoning is manifested prominently in American architect Daniel H. Burnham and other architects' bold attempts to reimagine the American metropolis. Daniel H. Burnham called to "make big plans" in the 1909 visionary Plan of Chicago and encouraged to "aim high in hope and work" (Hines 1988). Furthermore, the ambition in landscape architecture is illustrated by landscape urbanism theories, among which inherent outward-looking and seeking connections with a wider context were supported by landscape urbanists (Thompson 2012). With the landscape ambition, organic parks at the metropolitan scale intend to integrate their interior and exterior resources and spaces as much as possible while establishing more connections and interactions with regional surroundings. In this sense, organic parks are acknowledged as extensive landscapes rather than close and isolated parks.

Third, the extensive dimension of organic parks is analyzed regarding the landscape-based horizontal development in North American cities as well as spatial morphology in a landscape-ecological sense. In 1996, Richard T. T. Forman proposed the "ideal patch park shape" in combination with ecological function. He believed that the "optimum shape for a patch (park) is generally 'spaceship shaped' with a rounded core for the protection of resources, plus some curvilinear boundaries and a few fingers for species dispersal" (Forman et al. 1996). As the ideal model is transferred into organic parks, the "core area" guarantees a large field against spatial segmentation and fragmentation; the core area is beneficial to the ecological reservation of natural resources and biodiversity; the "fingers" could be regarded as ecological corridors for regional linkages to other landscape systems and species movement. Essentially, the ecological model could deepen our understanding of the organic park spatial morphology at the extensive level, as well as the ecological functions that organic parks take on.

The above proposal of ideal morphology has embodied the research progress in the field of landscape ecology since the 1980s. In 1986, Richard T. T. Forman and Michel Godron offered a new cognition of terminologies for analyzing ecological systems. In the book *Landscape Ecology*, they specifically investigated several essential spatial patterns, such as patches, edges, corridors, and mosaics, as well as these patterns' influence on the flows of organisms, materials, and energy that occur across landscapes (Hill 2001).

On the basis of this advanced knowledge, North American landscape architects tried to change the angle of view to search spatial relationships between parks and urban context, enlarge parks' contact interfaces, and even intentionally shape different spatial structures to satisfy natural and human demands. For instance, with the completion of Freshkills Park, total open space in the system of greenbelt on Staten Island will be increased to 30 percent (NYCDPR 2006). On this basis, the limits of Freshkills Park have more various interactions with the nearby area through which some relationships between parks and the surrounding area are created. In fact, the park includes five main areas: the Confluence, North Park, South Park, East Park and West Park. According to the "ideal patch park shape", the Confluence seems to be the core area of this ecological model. Other areas can be seen as containing the figures as ecological corridors and interactions with adjacent areas. Each area has a distinct character and programming approach to shape diverse connections between people and the reclaimed land, based on the social and ecological principles of openness and extensiveness. On the one hand, the park creates opportunities for large-scale recreational activities and programs that are unique to the city, such as community events, birding, public art, outdoor dining and extreme sports. Moreover, people see the opportunity to create extensive pathways and trails for walking, running, organized marathons, bicycling and horseback riding. On the other hand, a limited system of ecologically sensitive park roadways is built to optimize local and regional access to the park and reduce local traffic congestion.

- Inclusive Dimension

Large size "affords distinct opportunities that are otherwise impossible in the compressed urban and public spaces of cities, allowing instead significant space for urban wilds and protected nature reserves alongside extensive leisure and recreational amenities" (Corner 2009). The inclusive dimension represents the ability of housing elements and relations in both urban social and ecological systems. It demonstrates that the advantages of organic parks are mostly enjoyed in suburban areas. The inclusive elements may refer to the great variety of lifestyles, cultures, activities, programs, and events, among others, within urban social system. They may also refer to multiple habitats of woods, meadows, marshes, and water bodies within an urban ecological system. Between these two distinct systems, the inclusive relations in pairs are generally composed of "nature and culture", "art and science", "the natural and the artificial", and "the static and the dynamic" (Berrizbeitia 2007).

As aforementioned, these social and ecological elements, together with their relations, represent the multiplicity of social and natural concerns. They are also integrated into a large, contiguous, non-fragmented, yet heterogeneous park area. In the landscape-ecological viewpoint, the largeness, containing more heterogeneity and interconnectivity, is always associated with ecological structures and functions difficult to be altered "through habitat fragmentation, reduction and simplification, partial restoration, or even complete re-creation" (Lister 2007). Compared with a relatively smaller size, large size is considered an ecological advantage.

Ecological and Social Qualities

Compared with conventional parks, certain changes in the concept of organic parks have been revealed. From the qualitative perspective, more detailed content on the ecological and social qualities of organic parks will be formulated. With an increase in concern about ecological and social qualities, the two elements are incorporated as much as possible into organic parks to improve corresponding effectiveness. The most significant distinction for organic parks is to establish interactive social and natural relations in a dynamic fashion. Through the interactions, organic park qualities are reflected greatly and explained as the following five distinguishing features.

- Complexity
  The built complex systems in the adaptive processes of site transformation, open to the unpredictable future.

- Diversity
  This refers to *heterogeneity* in a landscape-ecological sense.

- Sustainability
  This refers to *resilience* in a landscape-ecological sense.

- Appropriation
  This refers to the social self-organization of spaces along the thread of programmatic indeterminacy for flexibility and multiple demands of users.

- Identity
  This refers to ecological identity, emphasizing the unfolding of ecological functionality or *performance* through space occurring over time.

These features imply the theoretical interplay of varying landscape-related disciplines. The most prominent cooperation comes from landscape and ecology because of the emergence of ecological principles since the 1980s. In the North American academe, the influence of ecology is profound, allowing organic parks to serve as an ecological-orientated paradigm.

- Complexity

Concerning complexity, the primary issue does not concern complexity theories, but its significance for urban landscapes.

> "Complexity theories, which today are considered the solution for the problem of creativity, offer a new creative view of the intellectual and the material world, linked to the evidence that completely new structures of order can arise, but are not predictable". (Poser 2008)

Complexity is introduced into the North American landscape architecture because of the creativity, as the core of North American urban landscapes, articulated as the cultural context of organic park emergence. The pursuit of creativity becomes the original cause to combine complexity with urban landscapes and organic parks. Thus, it forms a new cognition of "complexity inherent in landscapes" (Berrizbeitia 2001). Furthermore, the complexity of ecology and program is inherent in organic parks (Lister 2007), reflecting that organic parks possess "ecological and programmatic complexity" (Pollak 2007). The ecological complexity would present a nonlinear understanding of nature in the new perspective of *non-equilibrium* ecology based on dynamism. The programmatic complexity indicates a series of self-organized programs adaptive to the needs and desires of people in an unpredictable urban life.

Specifically, complexity refers to "organized complexity" in space and time containing a significant number of variables whose behavior cannot be considered random (Weaver 1948). Organized complexity reveals "the evolutionary development of nature in its nonlinear structural dynamics" (Poser 2008) when the linear, deterministic, closed, and stable view is not enough to describe the understanding of contemporary nature. Its definition tells two aspects of complexity.

First, organized complexity is composed of more than two closely connected parts that are dynamic and interacting, such as variables. The variables in organic parks contain time, urban nature, and urban life. These three variables produce an interplay, defined as natural and social processes. Second, these variables may perform certain distinct behaviors in complex systems:

> "Complex systems are interconnected networks of processes (or functions) and structures (or elements) whose behavior is generally described as non-linear, unpredictable, dynamic, and adaptive, and is characterized by the regular emergence of new phenomena and the ability to self-organize." (Lister 2007)

In large-park planning, complex systems are usually regarded as a dynamic framework constructed by planners and designers. The framework in space and time appears as "strategic organizations" and "dynamic infrastructures" (Czerniak 2001). It aims to establish dynamic interrelationships between processes and material structures.

The two characteristics, process and unpredictability, are interpreted separately to comprehensively analyze the complexity of organic parks.

The term "process" is first grasped as space occurring over time. It underscores the "formation of space through process" or processes as the "principal generators" of space-making (Wall and Dring 2015). The concept may highly depend on the new cognition of "the dynamic nature of the material itself" (Berrizbeitia 2007) which is a demand for design processes rather than a landscape's final form.

In retrospect, the understanding of process is produced primarily in the comprehension of contemporary cities. Early in the 1960s, Jane Jacobs, under the influence of the field of biology, contended that "like the life sciences" (Jacobs 1961), the processes and catalysts of processes are the essence of cities (ibid.). At the end of the twentieth century, the contemporary city is further explained as "a constant process of unfolding rather than a rigid reality" (Corner 1997). From American architect and theorist Stan Allen's viewpoint, the spatial process occurs in the living urban surface where the ongoing urbanization is further construed conceptually as "living societal and ecological processes" (Berger 2009). Consequently, the two processes of urbanization regarding urban society and nature become the leading directions of organic park processes.

However, the complexity largely leads to a process-based design approach applied in the planning and design of organic parks ascribed to the following three obvious advantages.

First, the process-based approach allows the (re)construction and (re)development of urban green open space to slow down. This is primarily induced by the transformation of most urban post-industrial sites that reject the eagerness for speedy success and the pursuit of instant benefit through accomplishing in an action. It takes time for the organic park growth, self-organization, and transformation; for the equipment of infrastructure, and the development and maturity of natural system; and for dwellers to perceive, find, and experience the changing process. People will be involved and extend more patience, concern, understanding, and support during the process. It is indeed "an interactive responsive network" (Czerniak 2001) that makes the growth of parks closely related to residents, groups, and communities, while forming their open interactions.

Second, it encourages professionals to re-consider the dominating role of planners and designers. According to Alissa North,

> "With the process design, the designer's role does not end with a traditional final master plan where landscape elements are fixed in space and time, but rather with a framework capable of guiding the evolution of the site toward a desired and continually relevant trajectory." (North 2012)

In the concept of organic parks, North American professionals intentionally choose another way out. They began to "guide or steer flows of matter and information" (Corner and Allen 2001) through the established framework per se, containing landscape elements, dynamic variables, and interactions in a complex field. Consequently, design initiatives are not simply "willful, subjective or formal approaches" (Corner 2009), and professionals are not easy to determine or predict outcomes.

Third, it suggests "an adaptive management approach in which the effects of interventions are monitored, adjustments are made, and new directions and configurations emerge" (Mertins 2001). It implies that organic parks are handed to the gradual development process, during which an adaptation would work. Adaptation or adaptive management frequently operate after obtaining a certain development and accumulation of parks over time. There is a required selective modification in open areas, infrastructure (such as transportation system), various ecosystems, and programs. There are also continuous monitoring and maintenance for various ecosystems, accompanied by the modification. In conclusion, a flexible arrangement adapting to changes is highlighted.

Besides the process, the idea of unpredictability may be influenced by French philosopher Henri Bergson's understanding of life in *Creative Evolution* in 1944. "The role of life is to inject some indetermination into matter" (Bergson 1944), which may offer infinite creativity of both biological and imaginative life. Within the scope of life, North American planners and designers realize that there is "a need to liberate life so that its fullest potentials may come into appearance" (Corner 1997). The potential in life may more or less demonstrate when some room is left for a flexible arrangement and adjustment during processes. The path of an emerging exploration may guide more North American professionals to pay attention to the unpredictability possibly employed in the planning and design of organic parks.

In the field of ecology, unpredictability is construed through *emergence*. The ecological understanding of "emergence" is articulated by American naturalist George Salt, who suggested that "it refers to a property of an ecological unit that is unpredictable" (Salt 1979). Its generation will be explained in the following part of *resilience* in the dynamic ecosystem model. The term has been adopted in practical large-park projects for designers to describe their concepts. For example, the emergent appeared as subject headings in the finalists' design schemes of Downsview Park competition: "Emergent Landscapes" by the Brown and Storey team and "Emergent Ecologies" by the Corner and Allen team. Nevertheless, the ecological mechanism of *emergence* was not their focal point for envisaging contemporary urban landscapes. The *emergence* indicates that both the maturity of ecological systems and "new forms and combinations of life" (Corner and Allen 2001) will emerge with the open-ended and complex ecosystems evolving toward an uncertain future.

Concurrently, unpredictability also symbolizes the aspect of social life. It "accounts for unpredictable urban life that might arise from the confluence of program with circulation, as well as for the outcome of participatory processes" (Czerniak 2001). Briefly, unpredictability refers to programmatic indeterminacy in organic parks that will be illustrated in social appropriation and through the practical project of Parc de la Villette by Rem Koolhaas/OMA in the following sections.

In conclusion, what has been discussed in this part is not the concrete mechanism of complexity alone, but its apparent features and significance for organic parks. From the creative perspective, complexity brings about new possibilities. From one perspective, it helps set up an additional logic or choice for concepts of contemporary large-scale urban parks with sufficient arguments. Its establishment is strictly opposed to the traditional concept of pastoral parks onto which landscape architects have placed too much energy, though such conceptualized understanding is concurrent virtually in planning and design. In other words, the emerging understanding based on complexity is not substituted radically for the traditional one, yet its existence plays an evolutionary role. From another perspective, it offers a new possibility for structures: self-organizing spatial processes with unpredictability.

- Diversity (Heterogeneity)

The diversity of organic parks is interpreted likewise in a landscape ecological sense. It refers precisely to the basic term *heterogeneity*, implying the differences and diversity of a cluster of ecosystems, developed by Richard T. T. Forman. The heterogeneity traces back to Richard T. T. Forman's ecological understanding of landscape. He explained the concept of landscape within the realm of landscape ecology as follows:

> " ... a heterogeneous land area composed of a cluster of interacting ecosystems that are repeated in similar form throughout. [ ... ] The definition also indicates that ecosystems in the cluster are interacting. Thus, animals, plants, water, mineral nutrients and energy are flowing from one ecosystem to another in the cluster. Each cluster is both a source and a sink for different moving objects." (Forman 1987)

This understanding fully reveals that the *heterogeneity* of land contributes to an essential shift from a model of a single ecosystem to multiple interconnected ecosystems in a dynamic fashion to respond to "an unceasing barrage of perturbations" (Worster 1993). As articulated in the inclusive dimension of organic parks, *heterogeneity* indicates multiple, interconnected ecological systems or habitats at various scales.

- Sustainability (Resilience)

The renewed understanding of large-park sustainability is grasped plainly through the concept of *resilience* in the landscape-ecological context. It differs from sustainability in a common sense that retains a pure state of balance and harmony at the sociocultural, ecological, and economic levels. The term *resilience* was developed by C. S. Holling in the mid-1980s. In landscape ecology, *resilience* displays the adaptive capacity and function of the living ecosystem, "the ability to recover from disturbance, to accommodate change, and to function in a state of health" (Lister 2007). The irresistible disturbance is mostly derived from certain external agents, such as "wind, fire, disease, insect outbreak, and drought" (Holling 2001), or human activities. Precisely, the transformative capacity permits organic parks to develop in a sustainable way.

The *resilience* is uncovered through C. S. Holling's dynamic model of ecosystem development in 1992. This model involves an emerging paradigm of the natural ecosystem, in which "dynamic equilibrium has substituted for an older idea—the steady-state 'balance of nature'" (Hill 2001). Its proposal argued that the field of landscape ecology "has moved away from a concern with stability, certainty, predictability and order in favor of a more contemporary understanding of dynamic systemic change and the related phenomena of uncertainty, adaptability, and resilience" (Lister 2015).

Holling's dynamic model demonstrates three properties: "potential" (or "wealth") that "determines the number of alternative options for future", "connectedness" (or controllability) that "determines the degree to which a system can control its destiny", and "resilience" that "determines how vulnerable the system is to unexpected disturbances and surprises that can exceed or break that control" (Holling 2001).

In the transferred two-dimensional plane, the cyclic model is decomposed into two loops, implying four functions of the living ecosystem: "exploration" (birth), "conservation" (growth), "release" (death, namely, "creative destruction" by American political economist Joseph Schumpeter in 1950, and "reorganization" (renewal) (ibid.). The two opposite phases unfold in a sequence. The first phase is growth and stability, making the transition from "exploitation" to "conservation" (namely, from "r" to "k"), in which a gradual accumulation occurs, and the properties of potential and connectedness (y and x axes) increase. The second one is the "back loop" of an adaptive circle, making the transition from "release" to "reorganization" (namely, from "$\Omega$" to "$\alpha$") (ibid.), when ecosystems occur discontinuously and change periodically, such as during disturbance. *Resilience* is involved in the two dimensions. It shrinks from "r" to "k" while expanding from "$\Omega$" to "$\alpha$". The latter transformation of change and invention is inherently unpredictable, such as the *emergence*.

In the mechanism of ecosystem development, the states of growth, prosperity, and stability in the first stage used to be the concerns of most planners and designers, who take them for granted as optimum at the ecological level. In many cases, these states constitute their simplified comprehension of sustainability in a landscape-ecological context when the second stage is not considered. This alternative frequently corresponds to the relatively elaborate maintenance and management of large-scale parks, followed by extra economic costs. The situation is not only deprived of the wilderness of nature with the quality of resilience but also difficult to apply to large-scale parks at the metropolitan level due to economic considerations.

- Appropriation

"The park should develop over time as users inscribe their own traces into its various surfaces and pathways" (Corner and Allen 2001). The word "inscribe" is employed to emphasize the users' huge and transformative effect on the large site of a park. The users' individual traces, mostly triggered by spontaneous actions, activities, and public events, may exert a visible or intangible force on the site transformation. This inscription may boil down to a kind of everyday, continuous, and individual appropriation in open, equal, multiple, and flexible ways. The appropriation is exposed to an individual's choice and freedom, involving a large extent of personal willingness and desires. It represents the self-organization of organic park space at the social level.

Regarding the development of organic parks, unpredictability reflecting the complexity is mostly concerned with the understanding of social appropriation. Landscape architect Wolfram Höfer pointed out that the aim of landscape work is not simply to accomplish fixed demands for the public, while specific situations and demands may naturally emerge in the diverse, creative, and uncertain urban daily life since the meaning of life is not merely satisfied to be arranged:

"This is not a predetermined purpose that would have to be followed by a people on the basis of its inherited character in order to fulfil requirements by nature and destiny at a specific location. Rather, the meaningful purpose emerges in the process of living in the landscape and is subject to change as part of that process." (Höfer and Trepl 2010)

- Identity

According to the profound influence of new ideas in landscape ecology, the cultural identity of large-scale urban parks without doubt tends to be organic. As articulated above, ecology in organic park concepts surely provides a useful analogy for complexity, diversity, and sustainability. The ecological metaphor for contemporary urban landscapes plays a major role in specific North American cultural contexts. This point has been explained in the second chapter of the 1990s North American analysis of the urban landscape.

The ecological identity of organic parks essentially relies on the unfolding of ecological functionality or performance through the space occurring over time. The performance is closely associated with the living urban surface. In Stan Allen's perception, an urban surface is not a flat lifeless plane but a thick section full of characteristics and behavior. "The surface in landscape is always distinguished by its material or performative characteristics. Precisely, its performative effects are the direct result of its material characteristics" (Corner and Allen 2001).

### 4.1.6. Practice: Projects of Organic Parks

Three projects are presented in chronological order to analyze multiple conceptions relevant to North American organic parks deeply and vividly: 55-hectare Parc de la Villette in Paris, France (1987); 231-hectare Downsview Park as a proposal in Toronto, Canada (1999); and 891-hectare Freshkills Park on Staten Island in New York, the United States (2001). The prominent reasons for choosing them lie in their embedded advanced landscape or ecological ideas and approaches contributing to organic parks. Simultaneously, the three case analyses also argue for the above key points of the qualitative qualification of organic parks.

Notably, these large-scale projects were initiated positively by international design competitions. Among the three projects, the Downsview and Freshkills projects are remarkable for the presence of landscape architects on established interdisciplinary teams of consultants, such as urban planners, architects, and landscape ecologists. The overarching role of architects in previous regimes of urban design and planning no longer being apparent is illustrated in the two former projects. Moreover, landscape architects and ecologists are increasingly engaged in organic park projects to update ideas based on intersecting disciplinary knowledge. In a landscape-ecological context, there is a difference between the less ecological Parc de la Villette and Downsview and Freshkills Parks strongly incorporating the ideas of ecology.

Meanwhile, North American landscape architects anticipate helping their discipline flourish through recent large-park projects. They believe that there is an interdependence with architecture and urban organization, and construction could be formed from the perspective of landscape. This situation could be summarized by the renaissance of landscape in Chapter 3 when North American professionals call for a critical readjustment to expand the scale and scope of the contemporary landscape.

Parc de la Villette

The 1982 competition for Parc de la Villette within the industrial periphery of Paris represents the beginning of the conceiving of "the urban park for the twenty-first century" (Tschumi 1987), as exhibited in Figure 30. In the field of landscape architecture, the large-scale park's provoking concepts laid the foundation for the rise of North American *landscape urbanism* and a new type of urban park with a broad and multicultural range of ideas.

**Figure 30.** Parc de la Villette, Bernard Tschumi Architects, 1987. Source: Photo by ©Danzi Wu 2015; used with permission.

For North American landscape architects, Bernard Tschumi architects' and Rem Koolhaas/OMA's proposals are considered to be the first examples and pioneers of *landscape urbanism* theories, and their submissions signaled a paradigm shift in contemporary parks (Waldheim 2006). Their remarkable views contributed to the further conceptions of large-scale urban parks, involving two levels: first, "programmatic indeterminacy" was introduced into contemporary landscape architecture (Koolhaas and Mau 1995); second, disrupting polarization in the North American critical perspective was emphatically proposed by Elizabeth Meyer in 1997.

- Programmatic Indeterminacy

The second-prize proposal by Koolhaas/OMA presented a conceptual approach to the landscape process, during which 'programmatic indeterminacy' was emphasized. In the metropolitan field, "orchestrating urban program as a landscape process" (Waldheim 2006) is regarded as an essential design strategy.

Since the 1970s, Koolhaas and his colleagues have continuously focused on and critically developed the role of the "program" in the making of projects (Wall 1999). Their proposal for Parc de la Villette is a strong confirmation of this. In their opinion, the idea of "program" is pushed toward more dynamic and productive ends, and the program is considered the engine of a project, driving the logic of form and organization while responding to the changing demands of society (ibid.). Their idea reflects the characteristics of openness and adaptability. Openness reveals that programs are no more strictly fixed or arranged in advance by designers. They may freely and flexibly take in potential possibilities from specific sites and users and are even open to an uncertain future. The dynamic process is adaptive to changing social demands. As Rem Koolhaas and Bruce Mau explained:

> "It is safe to predict that during the life of the park, the program will undergo constant change and adjustment. The more the park work, the more it will be in a perpetual state of revision. Its 'design' should therefore be the proposal of a method that combines architectural specificity with programmatic indeterminacy." (Koolhaas and Mau 1995)

The programmatic indeterminacy in OMA's scheme is embedded in their diagrammatic plan. The plan is composed of multilayered diagrams: "strips", "confetti" or point grids, "access and circulation", and "the final layer".

The parallel "strips" with a width of 60 meters could accommodate major programmatic categories across the site, such as theme gardens and playgrounds. According to Rem Koolhaas, they "create the maximum length of borders between the maximum number of programmatic components and will thereby guarantee the maximum permeability of each programmatic band and—through this interference—the maximum number of programmatic mutations" (ibid.).

"Confetti" is formed by small-scale elements on grid points, such as kiosks, playgrounds, and picnic areas. Regarding the desirable frequency, the distribution of these elements is mathematically built up.

"Access and circulation" are formed by the boulevard and promenade. The boulevard as a major axis connects large-scale architectural elements, and the promenade reaches specific areas. "The final layer" is a composition of the major elements, which are large-scale buildings such as museums and halls.

Rem Koolhaas/OMA described the multilayered diagrams as a "landscape of social instruments" where the quality of the project would derive from uses, juxtapositions, and adjacency of alternating programs over time (ibid.). His description reveals the interrelationship between established diagrams, a framework, and indeterminate social programs. The conceived framework bears programmatic changes and corresponding social demands over time. The landscape is viewed as the "suitable medium" (Waldheim 2006) for supporting their occurrences.

A visionary perspective on contemporary parks changed the course of OMA's idea in their proposal, which is combined with highly changeable and unpredictable characteristics of urban society. In this plan, the visionary perspective is reflected in the focus on a strategic organization or precisely a conceived framework instead of a specific form. Rem Koolhaas/OMA's Parc de la Villette concept assumes significance for further North American organic parks because he stirred imagination coming from treating the contemporary city in a dynamic way and urban life in an unpredictable way. This imagination is manifested as the idea of urban programmatic changes. The imagination of the city and life is incorporated into the understanding of contemporary parks. To sum up, this scheme may become the beginning of introducing an indeterminate factor in a social meaning into the concept of twenty-first-century parks, especially large ones.

- Disrupting Polarization

In Bernard Tschumi Architects' and Rem Koolhaas/OMA's proposals, disrupting polarization becomes the second level influencing the organic park conceptions. This point demonstrates the criticism toward binary thinking regarding organic parks. It also suggests the close connection between the park and the contemporary academe of North American landscape architecture in the critical thinking discussed previously.

In Bernard Tschumi Architects' competition-winning scheme (Figure 31), he purposely validated that it was possible to construct a complex architectural organization without resorting to traditional rules of composition, hierarchy, and order (Tschumi 1987). He intended to "encourage conflict over synthesis, fragmentation over unity, madness and play over careful management" (ibid.). His concept of spatial construction is usually influenced largely by *deconstruction*.

**Figure 31.** The competition-winning scheme, Bernard Tschumi Architects, 1982. Source: Image courtesy of ©Bernard Tschumi Architects; used with permission.

Alternatively, he uses a deconstructive approach to disrupt the clear polarizations or oppositions between culture and nature, urban and rural, and form and function. French philosopher Jacques Derrida, the founder of *deconstruction*, stated that the critical aim of applying a deconstructive approach is "not to reverse or replace the binary, but to derail the whole system, creating a space for ambiguity, difference, and playfulness" (Derrida 1967). The entire system established through overlapping points (the red enameled steel follies that support different cultural and leisure activities), lines (movements such as the promenade, alleys, and linkages), and surfaces (large areas for mass entertainment and open spaces) by Tschumi demonstrates the essence of *deconstruction*.

OMA's scheme also calls for eliminating the destructive polarization through the repetition of a similar horizontal structure, within which the built and vegetal material is arranged (Meyer 1997). The structure is reflected in the diagram 'strips'. Rem Koolhaas attempted to blur traditional boundaries between nature and artificiality through the structure. The structure underscores "modes of distribution," "the ways in which materials and elements are arranged, rather than the things themselves" (Choay 1985). His overall plan reveals that the built and vegetal as component elements are more or less difficult to be discerned. Such confusion exists because he employed a nonhierarchical planning strategy that does not rely on binary opposites. Elizabeth Meyer provided a concrete analysis of such confusion in 1997:

> "Their form and structure are not one of contrast, built versus vegetal, but of similarity. This repetition of alternating built and vegetal strips calls into question the oppositional nature of naturalness and artificiality. [ . . . ] This confusion of categories, wherein the vegetal can be artificial or human-made and the built can be scattered or natural, refers back to the traditions of nineteenth-century park and promenade design and addresses the 1982 competition brief's call for 'the contradictory requirement that the park be at once thoroughly natural and cultural' (Choay 1985)." (Meyer 1997)

Downsview Park

The second project is Downsview Park in Toronto. Given its ideas of planning and design in the competition schemes, it is almost the first organic park case acknowledged by North American professionals. The Downsview Park was a decommissioned military base and is now regarded as Toronto's "first major new park of the twenty-first century and an integral part of the city's attempt to intensify itself" (Glover 2001).

It is a crucial organic park case, in which the definition of a large-scale urban park in flux is encouraged to understand, and the transformation of the site is inaugurated, while remaining open to change and growth over time. These descriptions can be observed in the 1999 Downsview Park International Competition Brief, in which two of the five finalists' design schemes from the Tschumi and the OMA teams are mentioned. They demonstrate that the building of a large-scale park to offer a complex urban landscape requires designers from varying fields, including landscape architecture, graphic design, and ecology. Moreover, both of them are inclined to favor organizational "frameworks over form", considering that established frameworks in their proposals may "offer the possibility of both accommodating the three-stage, fifteen-year implementation process of the park, with its attendant public programming opportunities, and anticipating the transformability, emergence, and complexity of natural and cultural processes" (Czerniak 2001).

- Tschumi Team's "The Digital and The Coyote" Scheme

The most noticeable feature of Tschumi Team's proposal is the dynamic blending of the natural and the cultural through the construction of a framework. Their apparent interplay is displayed within interconnected landscape systems that flow spatially along the edges of the park.

The original land of Downsview competition is augmented, from about 21 hectares to 42 hectares. The augmentation might have been influenced by the landscape ambition of thinking beyond the given in large-scale park conception. Concurrently, its purpose lies in establishing more linkages between the Downsview Park system and other ecosystems in this wide region, involving two major ecological corridors: Don River System and Humber River System. In this sense, the park is positioned at the center of interconnected landscape systems on a regional scale.

On the basis of the existing layout, linear connections to the two ecosystems in the proposal are thus set up by appropriate extensions toward the surrounding landscape context. They are considered corridors shaped by regional woody vegetation. Undoubtedly, these wooded corridors will not only evoke an association of Greater Toronto Area's remarkable landscape element and character, which is woodland, but also assume an ecological function and role. They have established linkages for the movement of people, water, and wildlife species, explained in the part of "ideal patch park shape". The corridors as an essential spatial pattern are channels for exchanges of material, energy, and information between the park and surroundings in a landscape-ecological context.

Describing the scheme metaphorically as "The Digital and The Coyote" suggests an understanding of a complex park site on which there are two juxtaposed urban realities: digital culture and wild nature. For Tschumi, everything is "urban" in the twenty-first century, "even in the middle of the wilderness" (Czerniak 2001). This viewpoint primarily reflects a designer's attitude toward actively blending the cultural and the wild instead of adopting a binary separation and opposition. We can interpret the design proposal as a framework over time: dynamic integration of nature and culture.

The Tschumi team discovered the following design approach to handle the above two realities. They strived "to mix, to permeate one another in the most positive and fluid—liquid—manner" (ibid.). For this purpose, the first thing is to increase their interface by maximizing the presence and length of the park perimeter. The team introduces the concept of "Digits" with the characteristic of fluidity derived from a "fractal phenomenon of viscous fingering" (ibid.). The "Digits" direct the park's edges porous to admit its surroundings (Pollak 2001). The team makes the edges and interfaces within a fractal, fluid scope. Then, a distinct perimeter landscape is envisioned.

The park's framework is organized with the concept of spatial fluidity. It is composed of three superimposed conceptual elements: "Digits", "Spools", and "Screens". They are primary physical and spatial means for defining and activating the park. Each of them functions for stimulating space for continuous development over time instead of shaping differential, specific spaces, or fixed forms. They are definitely non-site spatial elements. Designers resort to the organized framework to conceive the large park morphology, with few referring to the physical geographical morphology. The latter move is increasingly viewed as a consideration of strategy. Tschumi purported that "conceiving of any large spatial organization begins with a strategy, never with a form", namely, "frameworks over form" (Czerniak 2001).

## - OMA Team's "Tree City" Scheme

The Tree City submitted by the OMA, Bruce Mau Design, and other entities won the international competition. Their proposal demonstrates a closer relationship between the conceived urban condition and large-scale park. The diagram employs a framework approach applied to the park, which is similar to that applied to their Parc de la Villette competition program.

Above all, the scheme is related to envisioning an urban condition. Bruce Mau, one of the park designers, stated that "to imagine a park presumes an urban condition" (Mau 2000). How designers conceive of the large-scale park reflects how they perceive the urban context. In the OMA team, Rem Koolhaas's viewpoint could guide their concept of the park. He suggested that landscape, the essential element of urban formation, could depict the urban condition as "a sparse, thin carpet of habitation. Its strongest contextual givens are vegetal and infrastructural: forest and roads (Koolhaas and Mau 1995). The core of Rem Koolhaas's idea corresponds to the above-uncovered thought of *landscape urbanism*, and landscape-based urbanization became their imagination of urban condition. In other words, OMA's proposal is rather sensitive to the 1990s North American urban landscape formulation of *landscape urbanism*. The concept of organic parks is connected tightly with the renewed theory of contemporary urban landscape in this paper.

Accordingly, the urban condition in the Tree City is understood ultimately by the designers as "low-density metropolitan life", catalyzed and realized by growing landscape elements, such as trees, quoted in Julia Czerniak's 2001 book *Downsview Park Toronto*:

"Trees rather than buildings will serve as the catalyst of urbanization. Vegetal clusters rather than new building complexes will provide the site's identity. An urban domain constituted by landscape elements, Tree City attempts to do more by building less, producing density with natural permeability, property development with perennial enrichment." (Czerniak 2001)

For the OMA team, the "low-density metropolitan life" exactly expressed the landscape scene described by Rem Koolhaas. It may indicate the generation of urban density and the conception of low density in the suburb of Toronto City. Downsview Park, located in the midst of one of the city's major potential suburban intensification areas, aims to become "a catalyst for suburban intensification" and bring about the anticipated population growth (Glover 2001). Guided by the construction of a large-scale park, an additional 8000 residents would live on this site. It is the process of producing urban density. However, what will be emphasized in the plan is the formation of low-density urban life through the distribution of trees and infrastructure. They are precisely the circular vegetal (or landscaped) clusters complemented with 1000 pathways.

Second, the diagram is formulated for process. The term diagram has been early employed by Rem Koolhaas for the construction of a large park framework in the 1982 Parc de la Villette competition. For OMA's designers, the virtue of the diagram in the large-scale park planning generally lies in its "vague specificity that permits future diversity" (Somol 2001).

For Downsview Park, his team further explored this framework approach, which is displayed explicitly in a way of distributing circular vegetal clusters. On the entire site, these clusters of varying sizes are vividly described as the planted seed for environmental expansion. They seem as if they were circular icons representing the park components and even "acted as programmatic and formal placeholders to be filled in appropriately over time" (North 2012). American architectural theorist Robert Somol revealed that the purpose of the diagram is to realize meaningful environmental expansion with the maturity of vegetation rather than to shape specific spatial forms. This vegetation will satisfy Tree City's emergent programs over time. In conclusion, the diagram is used for the process of park growth.

Freshkills Park

The third case of North American organic parks is Freshkills Park, which is among the most familiar projects of almost worldwide professionals in landscape architecture. In the 2001 "Freshkills Landfill to Landscape international design competition", Lifescape led by Field Operations became a winning entry.

Lifescape is "an infrastructural strategy of emergent colonization that stages various systems and sets in motion a diverse ecology of events and the complex organizations of forms" (Corner 2007). As revealed by James Corner, Freshkills Park is essentially positioned as an urban "organic infrastructure" (Marton 2010) in accordance with the 2010 new understanding of the twenty-first-century parks proposed by NYCDPR. Parks are a crucial component of the urban infrastructure that will help our city address the challenges of the twenty-first century (Bloomberg 2010), as stated by former New York City Mayor Michael R. Bloomberg. In North America, one of the challenges is how to greatly increase the ecological functionality of both contemporary cities and urban landscapes over time. In this sense, parks undoubtedly undertake the role of urban organic infrastructure.

In this context, the ambitious Freshkills Park project in the New York metropolitan region emerges at the right moment and becomes a convincing case on the aspect of constructing the twenty-first-century park. The anticipated goal of Freshkills Park in Field Operations' Lifescape Draft Master Plan is to "transform an industrial landscape into a state-of-the-art environmental preserve and innovative, contemporary urban park" (Field Operations 2006). Hence, the site transformation calls for combining advanced ecological restoration techniques with extraordinary settings for wildlife, active recreation, public art, and facilities for diverse activities and programs.

The two crucial points in this goal are the process of transformation organized into successional phases and a matrix as a conceptual approach to reconstitute "diverse life-forms and evolving ecologies" (Field Operations 2001).

- Process of Transformation

James Corner summarized that "lifescape is both a place and a process" (Corner 2005). He suggested that the process, "growing the park over time," is central to the project, because a large-scale site and its complexity could not be totally "designed" nor constructed overnight (ibid.). Hence, Freshkills Park calls for the process-orientated approach to construct facilities, cultivate native habitats, drive activities and programs, and finally realize the whole site transformation. This process would guide the site's development over the span of 30 years, during which there are generally four successive sequences of stages: seeding (the re-establishment of the original natural environment), infrastructure, programming, and adaptation. These stages surround the formative processes of four directions: circulation, surfaces, ecology, and program, or the concluded three new systems: circulation, habitat, and program.

Meanwhile, the whole plan for the ongoing organic park along the timeline is clearly through six principal implemented contents, based on "X"-scape. "X" represents "mound," "field," "open," "place," "event," and "life" (ibid.). Through the continuous accumulation of the prior five stages, the landfill would gradually develop into Lifescape, the theme of the Field Operations proposal. Specifically, the "X"-scape in various stages is formulated as follows.

- "Mound-scape"

In this stage, the Freshkills site is a closed landfill, without public access or amenity. This "engineering ground" (ibid.) is composed of the "mound-scape" together with other existing natural resources including creeks, wetlands, and open fields, such as grassland, meadow, and woodland. Four landfill mounds lend an unusual large-scale topographic character to Freshkills. They totally embrace 150 million tons of waste, occupying 45% or 69 hectares of the land. In conclusion, the "mound-scape" depicts the impressive site condition and its ecological challenge. Hence, it is crucial to offer a re-imagination of the huge open space with unique features of metropolitan location, openness, and ecology.

- "Field-scape"

In the first three years of the conceived process, there are primarily two steps toward the land transformation: remediating the soil and stabilizing the slope with the agricultural practice of "strip cropping" as an inexpensive and large-scale technique, and subsequent "propagation of plant communities" for emerging native habitats across Freshkills over time, concluded as "field-scape — manufacturing soil and habitat" (ibid.).

- "Open-scape"

Thereafter, the park is built as an urban open space by initiating access around the park and activity. Connecting Freshkills to the surrounding urban transportation system is the main approach to establish accessibility on a large scale. Urban activities could unfold with the solution to the access problem accordingly.

- "Place-scape"

The shape of a place begins in the first 10 years. Ground manipulation as the main content aims to generate earthwork and landform buildings for supporting park programs.

- "Event-scape"

In the next 10 years, the event-scape will occur after most of the facilities and infrastructures have already been well organized and the original natural environment has been re-established in the park place. Specifically, the first two of four stages, seeding and infrastructure, have been accomplished. Hence, the event-scape suggests the dynamic situation of "diversifying ecologies and uses" (ibid.), as well as the stages of programming and adaptation.

- "Life-scape"

Freshkills Park and new life would grow during its 30-year development. "Life" actually stands for the coexistence of wildlife and sociocultural life across "a mature biomatrix" (ibid.). In conclusion, the Lifescape proposes "a growth emergence from past and present conditions towards a new and unique future" (ibid.).

- Matrix as a Conceptual Approach

Freshkills Park's complex systems are conceived by citing a landscape-ecological concept of "matrix". As discussed above, the matrix is among the essential spatial patterns in the Richard T. T. Forman and Michel Godron's 1986 studies. The matrix is characterized by "porosity (or the density of patches), boundary shape, networks, and heterogeneity" (Pollak 2007). It "plays the dominant role in the functioning of the landscape, including the flows of energy, materials, and species" (Forman and Godron 1986). What will be reiterated is that the ecological matrix in a dynamic fashion not only assumes a leading role of containing and connecting habitats of diverse sizes and shapes to support heterogeneity but also guarantees the interactions and movements of all forces and agents.

Aside from its concept and ecological functionality, the matrix could be employed in the organic park planning and design owing to its holistic and multiple views. This point has been proposed by architect Linda Pollak:

> "In a design project, a matrix can support the construction of a kind of unity that does not rely on a single vision or overarching order to manage in creative and operational terms the interactions between multiple perspectives, scales, and types that attend the development of a complex urban ecological landscape." (Pollak 2007)

Moreover, the matrix is "an initial framework" from the analytical perspective (Prominski 2005), precisely through which a large park site could take form, evolve, and transform flexibly to accommodate the varying needs of a changing environment during the process. In Lifescape, the matrix was used as a conceptual approach to create a multi-layered and dynamic spatial framework. It comprises superimposed site layers past and present, involving existing systems and three new systems: habitat, circulation, and program. The framework is cast in view of the four stages of development: seeding, infrastructure, programming, and adaptation over a 30-year timeframe (Pollak 2007).

The superimposition of multi-layered structure applied in landscape analysis may originate from Ian McHarg in the 1960s, who pioneered the concept of ecological planning. It is a method of landscape analysis that has contributed to an understanding of the layering of different parameters in the design of a landscape (McHarg 1969). The method is frequently used, especially in geographic information systems. He overlaid maps of diverse natural and social factors to better understand the interaction of natural and social processes (ibid.). However, the maps shaped by deterministic geologic processes are relatively "closed" to interactive influences outside the local area.

The present-day organic park planners and designers inherited Ian McHarg's layering approach. Under the consideration of the open-ended exchanges of energy and information across urban landscapes, they further developed the dynamic matrix under the influence of new ecological ideas distinguished from Ian McHarg's approach. Notably, how interactive processes operate in space and time is visualized by the matrix.

Specifically, three coordinated conceptual diagrams; threads, clusters (or islands), and mats; constitute "an expansive green matrix of infinite horizons, interconnected ecosystems and pathways" (Field Operations 2006). These conceptual diagrams are collectively understood as "the agent of a fluid set of ecological systems, allowing the interaction of programmatic, cultural, and natural elements to create the complex, synthetic environment" (Pollak 2007). Hereto, Field Operations formulated them with the following different connotations:

> "Linear threads direct flows of water, energy and matter around the site, injecting new life into otherwise homogenous areas. Surface mats create a patch-like mosaic of mostly porous surfaces to provide self-sustainable coverage, erosion control and native habitat. Clusters of islands provide denser nests of protected habitat, seed source and program activity." (Field Operations 2001)

Influenced by James Corner's *cultural imagination*, the reasons why North American large-scale urban parks appear as the organic park model in former complex and contaminated industrial sites are analyzed in this chapter, followed by how these ecological ideas, terms, and conceptual landscape-ecological patterns are applied to organic park conceptions, based on James Corner targeting at design in landscape architecture. Complexity, resilience, processes, performance, and indeterminacy are essential concepts to understand organic parks in flux for social uses and ecologically sustainable effects.

In recent years, increasingly, large park-related projects have been brought to the public through international competitions, triggering far-ranging discussions and controversies over park ideas, approaches, and insightful design philosophy. Meanwhile, these discussions demonstrate that certain North American planners and designers tend to put their unique organic park conceptions into practice following the framework of *landscape urbanism*. The two points have been confirmed by Charles Waldheim in his 2006 *The Landscape Urbanism Reader* in which Downsview Park and Freshkills Park are taken as examples; "several recent international design competitions for the reuse of enormously scaled industrial sites in North American cities have used landscape as their primary medium" (Waldheim 2006). These large park-related projects are "representative of these trends and offer the most fully formed examples of landscape urbanism practices to date applied to the detritus the industrial city" (ibid.).

According to research, the growth in the number of North American organic parks challenges the way large-scale urban parks are defined when professionals began to criticize a mode of fixed thinking, also described as "Either-Or". This prompts a critical consideration of whether the German large-scale urban parks could be redefined without *post-industrial landscape parks* or *landscape parks* to describe the changing urban landscape. This vision has been mentioned in the introduction. Consequently, the German model is also referred to as structuralistic parks with the idea of large thinking for the whole region. Urban regional transformation may be realized extensively with the German structuralistic parks as a strategy. Along this line, the structuralistic design paradigm of German large-scale urban parks will be analyzed regarding theory and practice.

## 4.2. Structuralistic Parks in Germany

Before the 1960s, European industrial wastelands with the traits of being filthy and unattractive were perceived as ruined areas of cities or black holes in the urban fabric. There are buildings, complexes, landscapes, and related equipment on these brownfields that serve as reminders of former or current industrial production processes (Li 2022a). For instance, Figure 32's depiction of the canyon landscape at the Welzow-Süd open-cast mine demonstrates the massive lignite reserves used in Germany. After the majority of the industrial buildings were destroyed, the entire area was in a ruined and abandoned state.

**Figure 32.** The Welzow-Süd open-cast mine resembles a scar across the whole region. Source: Photo by author.

In Germany, a number of carefully planned urban redevelopment projects, such as IBA Emscher Park and IBA Fürst-Pückler-Land in Lusatia (Lausitz), aimed to transform urban areas in a landscape-orientated way. For instance, under the Energy Heritage Route of Lusatian Industrial Culture of IBA Fürst-Pückler-Land program, the F60 Visitors' Mine project transformed the abandoned spoil conveyor bridge for open-cast mining into a driving force for tourism. Visitors can appreciate the allure of massive steel structures and the changing landscape, as shown in Figure 33.

**Figure 33.** A new industrial cultural landscape perceived and experienced by visitors in IBA Fürst-Pückler-Land program. Source: Photo by author.

These important initiatives for urban landscape renovation provided post- industrial regions with the social, ecological, economic, and creative impetus for structural change. The innovative, far-reaching projects demonstrated "the capability of space" (Zhu and Xu 2020) in site restoration, the power of landscape in social transition, and people's willingness to accept the industrial sublime. They also helped define the new urban cultural landscape. Landscape has been transformed from a noun to a verb, and as the American academic W. J. T. Mitchell's book *Landscape and Power* outlines, this has made it a cultural tool in the formation of social and national identities (Mitchell 2002). It is important to note that this updated understanding of contemporary landscape combined with *critical thinking* has gained traction in North American and European landscape architecture, largely catalyzing the development of expansive urban parks in these developed regions.

The people who are the first to recognize the possibilities of spaces will drive the transformation, as evidenced by the history of landscape architecture. One of the most important ways that shifting green open spaces were clarified in a dynamic way during the change was through the large-scale urban parks in Germany built on former industrial sites that incorporated vital landscape ideas and techniques.

Germany is also interested in the examination of cross-cultural landscapes because of its decades-long accumulation of post-industrial landscape theories and practices. In terms of landscape regeneration and wasteland conversion research conducted at the European level, Germany has emerged as the leading participant. Latz + Partner, Atelier LOIDL, Planungsbüro DTP, etc., who dedicated themselves to post-industrial landscapes and established examples in the planning and design of urban parks coupled with intricate landscape surroundings, had a significant impact on its park design paradigm. The majority of the theoretical framework for the German post-industrial landscape, in other words, has been organized by Latz + Partner through project work and landscape observation over the years. Other Germans, Germans abroad (such as Peter Latz's projected landscapes in Israel), and foreigners working on local German projects (such as Gilles Vexlard in Riem, München) also use the German large-scale urban parks with the contextualistic–*structuralistic approach*.

This section will present the German structuralistic parks from the perspectives of vision, conception, transformation, approach, qualification and practice. In parallel with the creative cultural condition of organic parks, the German large-scale urban parks will be analyzed in the cultural setting of urban regional transformation. Either regarded as a cultural landscape or a strategy for site transformation in urban renewal, the German park model reflects the cultural contextualization through *critical structuralism*. Additionally, a qualitative analysis is established to qualify structuralisitic parks. Five characteristics were applied to the analysis in the same process as in North America, but with entirely different interpretations.

### 4.2.1. Vision: Parks in Urban Regions

The discussion of structuralistic parks and the reflection of the term "park" in the profession of German landscape architecture are intimately intertwined. To help us understand what a park looks like in the twenty-first century, Peter Latz presented his insightful views:

"The 21st-century park is the acceptance of urbanity. This characteristic makes the park fundamentally different to its predecessors, which sought to screen and exclude urbanity. Accepting urbanity means consenting to a new set of rules or at least combining old and new ones. Public gardens or parks have been aestheticising agricultural or horticultural patterns for a long time. This holds true not only for the English garden but also for Renaissance gardens and their successors. Two of the sub-areas in the Landscape Park follow a traditional model, whereas in most other parts a new type of park has been developed that treats aspects of environmental renewal on a par with aspects of use." (Latz 2016b)

If the park must be considered from a historical vantage point, Peter Latz noted when Landscape Park Duisburg-Nord was being constructed: "This is to become a historical park, but the history starts now and goes forward as well as backwards" (Latz 1993). It means a new chapter in the history of park development is opened by the creation of large-scale urban parks on former industrial sites. More specifically, German large-scale urban parks are known as an "unconventional 'park'" (Latz 2008b). According to the American landscape and architectural historian and critic Marc Treib, these parks "primarily suggest a landscape and secondly a park" (Treib 2009), though the name word "park" appears to be remarkable. Specifically, a park has little in common with the English landscape parks from the eighteenth century while a landscape refers to the valuable urban cultural landscape. The twenty-first-century park represents a new landscape type, specifically *urban-industrial nature* (German: *urban-industrielle Natur*), that is directly tied to the finding of a new sort of nature in the industrial ruins (Kowarik 1992).

There are various types of nature in landscape architecture, ranging from the *first nature* that is provided, the *second nature* that is modified, the *third nature* that is found in gardens that are meant to evoke pastoral ideals, and the *fourth nature* that is found on abandoned sites that is *urban-industrial nature*. Meanwhile, since the 1960s, the acknowledgement of *urban-industrial nature* has been supported by the modern idea of *urban nature* (German: *Stadtnatur*), which emerged from urban ecology. *Urban-industrial nature* is likely to be accepted gradually as increasingly accessible urban natural environments become more widely noticed, enjoyed, and recognized by the general public. A crucial part of urban nature is *urban-industrial nature*, which can display the vivacity, wildness, and ecological restoration that nature possesses.

So, first and foremost, the many meanings, purposes and importance of *urban nature* in *urban regions* should be explained in the context of the German urban landscape, according to the following four crucial points of view.

- In the field of German landscape architecture, urban planning, and urban ecology, *urban nature* is a notion that is an interwoven concept. The development of the spontaneous cognition of *urban nature*, as shown in Figure 35, occurred as a result of a regular public survey on *urban nature* (Küchler-Krischu et al. 2016). *Urban nature* has become an essential concept for raising urban residents' awareness of nature and for creating a better future for urban living (ibid.). In all facets of urban life, people are able to perceive various types of urban green open spaces, such as the Munich English Garden, one of the world's biggest urban parks in the world, and its Beer Gardens (German: Biergarten), which serve food and beer next to the Chinese Tower in Figure 36, as well as the natural or nearly natural spatial atmosphere in the *urban region*. As seen in Figure 37, the River Isar, one of Munich's most important greenways, transforms a canal-like waterway into an untamed mountain river through the process of re-naturalization. It creates lush, new homes for both plants and animals, depicts the cycle of life, and follows city inhabitants in Munich through all the stages of their lives.

- According to *Master Plan Stadtnatur*, *urban nature* is broadly defined as all ecosystems that are important for biodiversity (BMU 2019). *Urban nature* is a comprehensive design of natural systems with biodiversity conservation as its core, in conjunction with regional green infrastructure planning, allowing nature to do its work and guaranteeing the urban ecological balance, with the protection and wise use of natural resources as a precondition. *Urban nature* contributes to ecosystem restoration, optimizes the provision of ecological services, offers nature-based solutions (NbS) to numerous issues including climate change and the loss of biodiversity for the creation of a green, resilient city, and ultimately leads to human well-being through planning and management at various scales.

- *Urban nature* is regarded as a tool for fostering a stronger bond between people and the natural world. It not only assists people in understanding urban space and the natural world as they mature, but it also significantly contributes to the growth of a sense of the natural world and encourages both physical and mental health. People's regular proximity to nature can reduce stress and improve concentration and performance. In fact, a growing number of teenagers view *urban nature* as a relevant learning space where they can explore to develop their self-awareness and sense of social responsibility, as well as their creativity, and show off certain social and athletic skills. They do this in nature experience areas, urban parks, community gardens, and residential green spaces such as those where they observe natural elements including soil, water, plants, and animals (Li 2022b). Urban nature has great potential to contribute a satisfactory and mutually respectful relationship between people and nature in a dynamic manner.

- *Urban nature* is viewed as a link to strengthen social cohesiveness and a sense of belonging in German society. It acts as a social glue that can renew city residents' faith in collaboration and mutual integration, as evidenced by the green spaces where urban life and nature are inextricably entwined. People establish a solid connection between themselves, the city, and nature through landscape management and upkeep, community development, etc. For instance, at the expansive Berlin Tempelhofer Feld, one of the biggest inner-city parks in the world that was once an abandoned airfield, city dwellers participated in numerous leisure activities and gardening projects. People enjoy open urban green spaces and are looking for their original sense of community while they work to create *urban nature*, as shown in Figure 34.

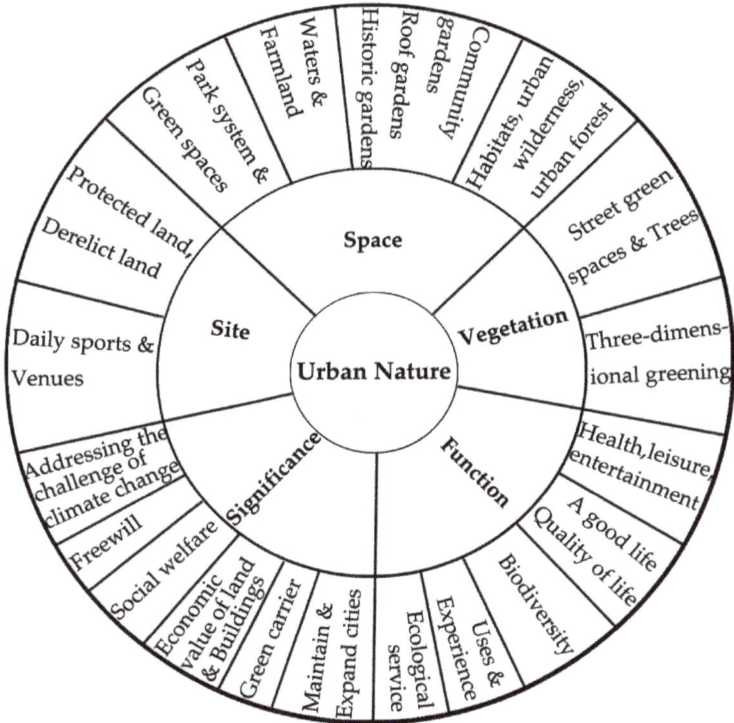


**Figure 34.** The system of urban nature perceived spontaneously by the public. Source: Figure by author.

**Figure 35.** English Garden as one of the popular green open spaces in urban daily life providing an oasis in the center of Munich, Fredrich Ludwig von Sckell, 1789. Source: Photo by author.

**Figure 36.** Munich River Isar offering diverse spaces for social interation and daily activities in an intertwined system of blue and green spaces. Source: Photo by author.

**Figure 37.** Berlin Tempelhofer Feld, McGregor Coxall, 2010. Source: Photo by author.

Furthermore, the idea of *urban-industrial nature* can be understood based on a thorough understanding of *urban nature*. It has aided in the growth of urban ecology's biodiversity. As a result, urban nature conservation has changed in both type and scope. Landscape architects have started to let nature take control of the land using the wilderness characteristics of the site after ecologists' research on diverse urban natural habitats drew their attention to post-industrial sites (Figure 38). By allowing species to compete and thrive on their own, they have begun to provide diverse habitats for plants and animals in natural succession. Large-scale urban parks on German post-industrial sites receive new meaning and life from the *urban-industrial nature*.

**Figure 38.** Track Wilderness in Berlin's Park am Gleisdreieck, Atelier LOIDL, 2011.
Source: Photo by author.

Based on the evolving understanding of the park, structuralist parks emerged in the cultural context of urban regional transformation and were affected by the concept of *urban region* that was explained by European scholars. German large-scale urban parks, as interconnected urban green open spaces, reversed the old dictum that *towns devour nature.* City and nature are more syncretic and penetrative with each other.

On the one hand, in the transformation and redevelopment of urban regions, the landscape became a key issue (Gailing 2005). This was interpreted in Chapter 3 when the situation from which the German concept of large-scale urban parks emerged was examined. Based on this point, the German structuralistic parks in urban regions are regarded as an essential strategy to realize a series of fundamental changes. They also consider socio-culture, ecology, and open space policy. A considerable amount of attention has been offered to the qualities and potential of landscape in the post-industrial site transformation.

On the other side, from the perspective of dissolved urban structure, the German structuralist parks are intertwined inextricably with the concept of the *urban region* through the European analyses. In the opinion of German scholar and regional planner Ludger Gailing, urban regions and their urban landscapes are shaped by:

"Spatial trends like urban expansion and urban sprawl, and the fragmentation of open space by the construction of infrastructure networks and the consequent ecological problems and degradation of landscape aesthetics." (ibid.)

Generally, these urban spatial trends, accompanied by regional spatial problems in society, ecology, and aesthetics, become the significant regional background for park construction and development. They lead the German structuralist parks to be closely connected and improve corresponding spatial qualities and people's life. They are naturally associated with the urban system, no more than an isolated park system. This description reflects the integration of structuralist parks into the wider urban regions in a holistic view.

Therefore, it is crucial for planners and designers to manage the complex interrelations between structuralistic parks and surrounding environment in urban regions and build their connections immensely. From this point, the holistic perspective is required, and the *structuralistic approach* is effective. The Landscape Park Duisburg-Nord project in this section will particularly explain the establishment of interconnections between the external and the internal, under the consideration of line of sight, horizon, transportation system for accessibility, and local special landscape image and elements. In a word, the German structuralistic parks could largely combine the region and the local.

### 4.2.2. Conception: Parks as Cultural Landscapes

It has always been crucial for German landscape architecture to focus on the perspective of the cultural landscape (German: Kulturlandschaft). German large-scale urban parks are thus viewed as an essential urban cultural landscape.

The concept of cultural landscape requires examination. In the early twentieth century, German geographer Otto Schlüter first defined this concept formally as an academic term (Martin et al. 1981). The geographical point of view essentially lays a foundation for analyzing the German cultural landscape regardless of its various understanding, debates, and subsequent development. From the geographical perspective, land, people, and their interaction are all treated in an objectivistic view. The cultural landscape is regarded as a physical object related to people's quality of life (Kirchhoff and Trepl 2009).

However, the perception of the cultural landscape was built much earlier than its academic definition. It emerged as a harmonious "unity of land and people" (Riehl 1851) ideally interpreted as *Heimat* (the place where people grow up), suggesting that homeland "evolved through the interactions of adapting to nature and cultivating it" (Hauck and Czechowski 2015) in Figure 39. Tracing back the original meaning of the cultural landscape, German biologist Wolfgang Haber stated:

**Figure 39.** German cultural landscape revealing the harmonious relationship between people and land. Source: Photo by author.

"When hunters and gatherers settled down to become farmers, they cultivated (wild) nature, thus founding agrarian culture, agriculture, by transforming the nature into cultivated land [ ... ]. It was from this cultivated landscape that the human environment developed at the expense of nature." (Haber 2010)

Briefly, the term *second nature* suggests that the original understanding of the cultural landscape, which connects the agricultural or pre-industrial society, and cultivated nature are represented in the idealized interpretation.

Moreover, the cultural landscape is generally developed with the technical and societal development and cultural advancement. The emergence of urban culture induced by the production surpluses of agriculture marks the transition into the industrial age (ibid.). During the 1960s and the 1970s, German society began to enter into a new stage, post-industrial society with the rapid expansion of social production and life, prompting the cultural landscape to evolve into another meaning. The new relationships between nature, the landscape, and people have arisen.

For instance, the German IBA Fürst-Pückler-Land established 30 avant-garde projects between 2000 and 2010, with the Geopark Muskau Coal Crescent being one of them. Figure 40 illustrates how this project, "Revealing the Landscape in Transition," intends to showcase and document the relationship between natural processes and human activities while also protecting this cultural landscape across national boundaries.

**Figure 40.** The IBA Fürst-Pückler-Land project of Geopark Muskau Coal Crescent. Source: Photo by author.

Urban industrial site discoveries encourage a continuing understanding of the cultural landscape. There are not many smooth transitions. The transition to a post-industrial society's structural shift was marred by disorder, confusion, and cracks. Since the 1960s, botanists, environmentalists, and landscape architects have found and studied the *urban-industrial nature*, which is the unique nature of urban industrial sites. In the 1980s, botanists began making observations in abandoned areas, which led to the nature of urban ecology (German: Stadtökologie). According to Dettmar and Ganser (1999), the phrase *"nature of the fourth kind"* refers to a difference between the cultivated nature in the traditional cultural landscape and the chaotic nature found in urban areas. The significance of this new interpretation of nature lies in starting to associate old industrial sites with a particular nature.

A wild and rebellious green, for instance, predominated the post-industrial site. Figure 41 shows the Landscape Park Duisburg-Nord, a nature of German landscape architects that captures the beauty of an *urban-industrial nature* reflected in the wild plants (including colorful butterfly brush) on a disused railway track. Peter Latz also highlights the new era, stating that "the time for a new understanding of nature has come" (Latz 1993).

**Figure 41.** Renaturated former railways in Park Duisburg-Nord. Source: Photo by ©Luca Maria Francesco Fabris; used with permission.

Botanists made investigations of abandoned areas and found a potential for new species to survive in the shadows of spoil heaps and head frames, which greatly contributed to the new understanding of nature (Siemer and Stottrop 2010). Since there is now a more modern understanding of nature, the relationship between cities and the natural world has changed. Within the context of the changing cultural landscape, wild nature or urban wilderness is widely recognized.

For instance, the IBA Fürst-Pückler-Land has resulted in significant alteration of the landscape and structural change in the mining region of Lusatia. The Terraces on the banks of the developing Lake IIse in the former mining town of Grossräschen-Süd welcome locals and visitors to experience the new wilderness and the art of structural engineering after the pits were turned into lakes. Figure 42 illustrates the role of the new landscape as a major motif in industrial culture. Large-scale urban parks incorporate the impression of wild nature. "Park soll öffentlicher Stadtraum und gleichzeitig Wildnis sein, in der natürliche Reize wirken können," said Peter Latz (Latz 2012). This suggests that a park should be a wilderness area with urban open space that functions through natural stimulation. A new type of urban cultural landscape is created by combining old industrial sites with wilderness to remodeled disused areas as structuralistic parks.

**Figure 42.** The urban wildness experienced by visitors in the mining region of Lusatia through IBA Fürst-Pückler-Land. Source: Photo by author.

In a way to integrate people's experience of nature with the aesthetics of wilderness, which is a temperament that cannot be accommodated by conventional park aesthetics, *urban-industrial nature* has been developed. With the emergence of a landscape design strategy that juxtaposes the sublime nature of industrial heritage with natural processes, the German landscape is no longer dedicated to the aesthetic feature of natural perfection but rather to the transformation of social aesthetic values on the post-industrial site.

Additionally, the repositioning of nature aided German landscape architects in regaining "its original roots" and "a new self-confidence in the profession" (Kühn 2013). Because design incorporates knowledge and data into its decision-making processes, it has the potential to "abstract by means of artistic intervention and edit incoming information down to the absolutely essential" (ibid.). This is where the design originates. After years of ecologically influenced ideologization in the 1970s and 1980s and the disproportionate emphasis on preservation and conservation, contemporary landscape architecture has reclaimed its freedom to design, according to Udo Weilacher (Weilacher 2005).

In conclusion, the analysis of the German cultural landscape suggests that its understanding is inseparable from the perceived nature in a specific society. Today's urban cultural landscape is naturally connected to the urban-industrial nature in German post-industrial society.

### 4.2.3. Transformation: Parks from Derelict Industrial Sites

There is a precondition of structuralistic parks on post-industrial sites for transformation, which is stated by Peter Latz as "a calm acceptance of the (industrial) structures" (Latz 2004) in the philosophy of "accepting a fragmented world" (Latz 2003):

> "Accepting the materials found on site, without placing them in traditional categories like beautiful or not beautiful, but just looking at whether they could fit in with the language system or not." (ibid.)

He explained that "our new conceptions must design landscape along with both accepted and disturbing elements, both harmonious and interrupting ones. The result is a metamorphosis of landscape without destroying existing features, an archetypal dialogue between the tame and the wild" (ibid.).

The concept of Peter Latz changed how an area that was once heavily industrialized was transformed. The transformation does not entail changing the current structures from a disorganized and fragmented park image to a harmonious one. This is because such a transition would drastically alter the physical properties of sites and wipe out practically all site-specific *information*, including local memory and history. For instance, a section of the Berlin Wall that concealed intricate historical information has been maintained, allowing people to preserve their historical recollections of the city and trace its development. Figure 43 shows a planned construction for the wall that was constructed with *minimal intervention* for a continuous route for public space.

**Figure 43.** The complex historical information segment concealed in the Berlin Wall. Source: Photo by author.

Peter Latz observed that the structuralistic park goes beyond a harmonious representation of painted landscapes in a perfect setting. Acceptance, protection, and prudent utilization of existing industrial structures and elements are important considerations during the transformation. Figure 44 illustrates how buildings on the Duisburg-Nord post-industrial site have been accepted and preserved. Structuralistic parks can be seen as a type of historical object with layers of different *information*.

**Figure 44.** Landscape Park Duisburg-Nord interpreted as a kind of historical object with multiple layers of information. Source: Photo by author.

Post-industrial landscapes are probably more common in China according to Peter Latz's philosophy of quiet acceptance. Figure 45 shows how Beijing Shougang Industrial Heritage Park, a sizable urban park on a former industrial site, exemplifies his idea, despite the fact that, from the perspective of urban nature, the wilderness on this former industrial site has certain sensitive, well-kept distinguishing characteristics. Consequently, there are two bullets that could aid in forming some shared perceptions.

- In this vast, intricate system of the post-industrial landscape, there always seems to be a conflict between preserving the historical, multilayered elements as best as possible and modifying them to foster new values.
- The role of a landscape prototype with essential principles may be understood by designers and academics as they uncover some connections and influences that exist in design. This may stimulate them to dig deeper and investigate cross-cultural ideas and approaches to landscape planning and design as well as their adaptive applications.

**Figure 45.** Shougang Industrial Heritage Park in Beijing, probably influenced by Peter Latz's philosophy in the process of reconsidering the transformation of post-industrial sites in China. Source: Photo by author.

Most importantly, accepting the materials found on site actually elicits a core concept of *information* in the site transformation. Peter Latz pointed out the significance of *information* through an example of Duisburg-Nord. He said, "wenn wir das einfach alles abreißen und für das Restgeld noch etwas Neues machen, dann bekommen wir eine so geringe Informationsdichte pro Quadratmeter, dass das nur langweilig werden kann" (Latz 2016a). These words signify that they would obtain little *information* per square meter, which would make the site boring if they just tore everything down and built something new.

According to André Corboz's theory of territory, "territory as palimpsest" (2001), which alludes to the archeological notion of layer formations and depicts a socially appropriated space and its inscriptions and attributes (Bucher 2013), the idea of *infor-*

*mation* is connected to the "palimpsest metaphor" of landscape (Clemmensen 2015). The palimpsest preserves and inscribes into the cultural landscape the rich social and cultural *information* and meaning that have been amassed in layers throughout the course of history. It is crucial for German landscape architects to have traces and *information*.

Moreover, he mentioned the *density of information* (German: *Informationsdichte*) to emphasize abundant and accumulated information in history, based on the post-industrial landscape as a historical object that can be explored in an interpretive analysis. In the spatial and temporal dimensions, all the meaningful *information* to landscape planning and design is considered abstractly by Peter Latz as *information flows* (German: *Informationsströme*) in chaos, which should be grasped and handled, and condensed and superimposed in both landscape elements and structures (Latz 2016a). The process for planners and designers is exactly the ongoing "decision-making process" (Latz 2008a). They require to find and discern "what force the existing objects already have, what density of information they already possess, and what density of information first has to be introduced into the project", as is quoted in Udo Weilacher's (1996) book *Between Landscape Architecture and Land Art*. Given these considerations, Peter Latz offered his general ideas about *information* in German:

> "Hier ganz abstrakt zu sagen: das sind Informationen, die wir einbinden —
> entweder indem wir darauf gucken oder Nutzungen suchen, dann heißt
> das, wir bleiben in der Historie, gehen aber nicht zurück, sondern in die
> andere Richtung: nach vorne." (Latz 2016a)

He concluded the understanding of *information* integrated either through looking at it or looking for uses. Then, that means we stay in history; however, we do not go back, but in the other direction: forward. In *information processing* (German: *Informationsverarbeitung*), planners and designers may obtain a large amount of *information*, among which they explore its significance of uses and find the elements for uses in the landscape. These semantic and pragmatic aspects are Peter Latz's two of three levels of *information processing*: "Die Bedeutung zu nutzen — das ist die Semantik" and "Die Elemente zu finden — das ist die Pragmatik" (ibid.). Moreover, Peter Latz reminded professionals to be concerned with the role of traces of history in *information processing*, which not only directs to going backward but also opens the possibility to move forward.

In addition to the above two levels, Peter Latz's *information processing* also embraces the third level of *syntax*, which helps him to benefit from the chaos and is closely related to structures (ibid.). To analyze the relationship between *information* and *syntax*, he stated from the linguistic perspective:

> "The language of things and the way things are combined create informa-
> tion that is linguistic in character. [ ... ] If they are to acquire this linguistic
> character, they need everything that language constitutes: they require a
> diversity of accurate terms and a strong syntax". (Weilacher 2008)

The *syntax* or *syntactical* design concept can be traced back to Latz + Partner's Saarbrücken Hafeninsel planning. Its *syntactical* concept is "intended to be acquired with a minimum of interventions, including the existing ruderal vegetation and deliberately work with the information levels available on site" (ibid.), as exhibited in Figure 46. Instead of "giving the Hafeninsel a superficial facelift and transforming it into a neoclassical picture-book park," the *syntactical* design "was not just to ensure a viable basic structure and thus the rhythm of the park but also to give the landscape a voice by linking up what is already there with new design elements" (ibid.).

**Figure 46.** Bürgpark Hafeninsel's existing layers of site information in Saarbrücken kept for the syntactical design, Latz + Partner, 1999. Source: Photo by ©Qi Huang 2016; used with permission.

In this sense, Peter Latz's *syntax* of landscape shapes in the planning and design through the *information processing* of chaotic sites. The level of *syntax* also reflects a rational and critical perspective of the conventional parks. Therefore, Peter Latz's *syntactical* structures in German large-scale urban parks are analyzed and explored based on the concept of *information*, considering that landscape is "not the images, but the abstractions, schemata of information layers or single systems that are required for understanding structure" (ibid.). His understanding of structure elicits the following content of the *structuralistic approach* applied to German large-scale urban parks.

## 4.2.4. Approach: Structuralistic Approach

As mentioned in Chapter 2, *structuralism*, initially developed in structural linguistics, affected the field of architecture in the 1960s when there was a movement of *structuralism* (Peisl 2014). The movement, particularly in the Netherlands, was characterized by "rule based arrangement, congeneric spaces without hierarchies, flexible expandability and mutable floor plans" (ibid.). While criticizing the modernist ignorance of history and all purely functionalist, sectoral and strongly form-orientated approaches (Weilacher 2009), Dutch structuralist and architect Hermann Hertzberger offered the following important principles of analysis and design of the *structuralism*, quoted in Arnulf Lüchinger's 1981 book *Structuralism in Architecture and Urban Planning*, and Udo Weilacher's 2014 paper "Structuralism in der Landschaftsarchitektur" in German:

"Jede Lösung an irgendeinem Ort und zu verschiedener Zeit ist eine Interpretation des Archetypischen[ ... ]. Wir können nur etwas Neues schaffen im Sinne einer anderen Interpretation bestehender Bilder, diese neu bewerten und sie für unsere Situation geeignet machen[ … ]. Entwerfen kann nichts anderes sein als Fortbauen auf dem Darunterliegenden und es sozusagen verbauen. Der Gedanke, jemals von einem unbeschriebenen wießen Blatt auszugehen und dieses unvermeidlich mit unwirklichen und sterilen Konstruktionen zu füllen, ist unsinnig und hat auch negative Folgen." (Hertzberger 1981; Weilacher 2014)

Hermann Hertzberger's statement revealed that every solution in any place and time is an interpretation of the archetype, in both general and particular situations. We can only create something new regarding a different interpretation of existing images, re-evaluate them and make them suitable for our situation. Design is likely to be constructed within cultural interpretations based on what is understood as existence underlying on site. He believed that the concept of design will not be derived from a blank sheet of paper, and inevitably filling it with unreal and sterile construction is nonsensical and also has negative consequences.

His idea of architectural *structuralism* was proposed in the context of critique and restraint of the modern functionalist idea employed in European cities. This background has been stated along with the explanation of the critical reconstruction of the contemporary city in Chapter 3. Hermann Hertzberger's analysis of the relationship between existence and design is reminiscent of Peter Latz's key concept of information to be found, handled, and introduced into the planning and design according to one's own interpretation. One of the claims about landscape, that "landscape is always subject to interpretation, and therefore each landscape is a part of culture" (Küster 2013), is also demonstrated by the essence of *structuralism*.

However, the *structuralistic approach* in German landscape architecture exceeds the structuralistic philosophy of architecture (Peisl 2014). Peter Latz expanded the meaning of *structuralism* by using certain theoretical aspects of the Dutch architectural movement. As aforementioned, he "found his way to structuralism via the writings of architects like Aldo van Eyck and Herman Hertzberger" (Weilacher 2008), and his "vocabulary identifies him as convinced exponent of structuralism in landscape architecture" (ibid.).

Following the aforesaid explanations, *critical structuralism* interpreted by Peter Latz as one of the critical approaches in research is highlighted in the context of *critical reconstruction* in German landscape architecture. This has been manifested in the development of different planning styles since the early 1980s. An example is *perspective incrementalism*. *Critical reconstruction* for Peter Latz is to cultivate a "fantastic landscape that will follow the industrial age that we have to address in a new and careful way", quoted in Udo Weilacher's 2008 book *Syntax of Landscape*. His "new and careful way" is exactly the same as the critical *structuralistic approach* in this study.

For this approach, the material and deep structures of a site for planning and design are especially valued by Peter Latz, quoted in the *Syntax of Landscape*:

"Yes, I am definitely certain at the bottom of me that in case of doubt, structure is more important than form. That is quite certainly correct, [ … ] structures are relatively unattractive at a first glance. They are not very exciting, they are usually neutral, something in the background, essentially, like the percussion in a band. The solo trumpet steals the show, but there is only a rhythm because the bass and drums create it. They both have to be there, however." (Weilacher 2008)

In this situation, Peter Latz applied the *structuralistic approach* to practices of many park projects. These parks offer his individual interpretations of *syntactical* structures, as analyzed by Udo Weilacher in German:

"Er interpretierte die vorgefundenen Strukturen neu, reicherte sie mit weiteren Bedeutungsebenen an, und verknüpfte Altes und Neues zu wachtums- , veränderungs- und wandlungsfähigen Landschaftsstrukturen, die von unterschiedlichsten Besuchern immer wieder neu zu lesen und individuell zu nutzen sind." (Weilacher 2009)

This analysis implies that the structures previously found on site are reinterpreted, enriched with further levels of significance, and integrated for their capacity for growth, adjustment, and transformation through linking the old and the new. Moreover, Peter Latz's understanding of structures is not just from an objective standpoint, though he respects abundant and accumulated information of sites in their history. Instead, his idea of structures essentially reflects an intersubjective perspective in philosophy that indicates "existing between conscious minds, shared by more than one conscious mind", as suggested by Oxford Dictionaries. This point illustrates that the structures can be interpreted differently and used individually by diverse visitors. Lucius Burckhardt's explanation in 1985 could appropriately clarify this perspective:

"Anyone designing a landscape must consider whether the meaning he is creating is such that it is comprehensible to other people, and also to people from other cultural backgrounds. In our pluralistic society, a design must be open to multiple interpretations." (Burckhardt and Brock 1985)

Moreover, two aspects of his *structuralistic approach* should be emphasized. Primarily, the structures are adopted to cultivate and develop diverse spaces for social appropriation in everyday life. According to Udo Weilacher, "Die freie Aneignungsfähigkeit von Strukturen durch die Schaffung polyvalenter Räumen zählte zu den zentralen Anliegen des Strukturalismus" (Weilacher 2009). This means that the free appropriation of structures through the creation of polyvalent spaces is one of the central concerns of *structuralism*. His statement reflects the shaping of diversity and difference of space, as well as the various appropriation of space together with forms of activities, programs, and events. This will be explained from the qualitative perspective to investigate the German structuralistic parks.

Another aspect of his *structuralistic approach* is associated with *minimal intervention* or *the smallest possible intervention*, which had been adopted by Peter Latz from Bernard Lassus and Lucius Burckhardt (Weilacher 2008). In the site transformation, Peter Latz explained that:

"It is more about taking items over in their totality and understanding their original functions[ ... ]. we want to keep them in their role and in their historical function, and sometimes invest the surviving building components

with new meaning that can stimulate new readings of existing material".
(ibid.)

The explanation is precisely the first two aspects of information processing at the semantic and pragmatic levels while expressing the principles that Peter Latz follows to realize *minimal intervention.*

In view of his principles, Peter Latz "rejected the notion of a 'master plan'" (Rosenberg 2007) and "never wanted to draw an overall plan" for his parks, such as the Landscape Park Duisburg-Nord (Weilacher 2008). Instead, he assumes his own landscape syntaxes: "a weaving of industrial structures" of informational layers, "abstract portrayal of formative basic elements of the landscape" and concerns "linking independent structural layers" (ibid.) to form superimposed landscape systems in which diverse and flexible spaces could develop with multiple social uses in everyday life. The unique syntaxes are concluded by Peter Latz as abstract structures, "overlay and connection of independent conceptual layers and structural elements" (ibid.). He selected this contextualistic–*structuralistic approach* to achieve his analyses and planning of sites and maintained substances in the industrial age to the greatest extent, so as to offer new interpretations of old industrial elements and remains.

In summary, the German contextualistic–*structuralistic approach* to its large-scale urban parks is expounded gradually in consideration of Peter Latz's interpretations under the influence of Dutch architectural *structuralism.* As among the critical approaches, it builds up unique landscape syntaxes or structures through the *information processing* instructive to the free development of diverse and different social spaces, with respect to his understanding of *minimal intervention* in the transformation of a former industrial site.

### 4.2.5. Qualification: Qualitative Analysis

The book analyzed the North American organic park design paradigm with both quantitative and qualitative methods. The quantitative perspective is reasonable for exploring that model based on its focus on the larger size closely related to the higher ecological *performance* and functionality. The logical proposition is rationally interpreted precisely in the North American cultural context. Meanwhile, certain basic explanations have been explained clearly concerning the qualitative method.

For German landscape architecture, the research on urban green open space has changed the method from the twentieth-century quantitative to present-day qualitative analysis (Schöbel 2006). The principal reason for this transformation lies in social and spatial changes that have caused a change in the arguments for the qualification of urban green open space (ibid.). With these changes, the quantitative method reflects the finiteness regarding the aspect of defining open space. Therefore, the German large-scale urban parks are discussed in the qualitative method accurately, as essential contemporary urban landscapes, in the face of methodological alteration.

The proposed five characteristics of complexity, diversity, sustainability, appropriation, and identity are stated here for "qualitatively developing space" (Weilacher 2008) of structuralistic parks in Germany, similar to the ecological and social qualitative analytical perspectives of the North American organic parks. These characteristics show how the discussion of large urban parks on post-industrial sites is influenced by culture and that there are significant differences in their perception and priorities.

- Complexity

  The site itself is the source of the complexity in landscape planning and design, which may be reflected in design goals. Therefore, awareness of and aptitude for reading the depth of the site read as the crucial work premise for complexity-oriented landscape planning and design (Zhu 2022). Planners and designers need to find out complex site *information*, including visible and invisible layers of *information* and elements, and then influence the layering of these elements embedded in each layer (Weilacher 2008). Professionals are required to deeply perceive a series of intricate and interweaved *information* on a specific site. They should also elaborately evaluate and purposefully choose the *information* and elements to build the large-scale urban park structure in combination with the professional's own views and ideas. It is the complex design process of blending existence with creativity. For instance, Figure 48 shows a system of elements in terms of slag heaps, railway, motorway, and road embankments that transformed the Duisburg-Nord site into hilly terrain.

  Alternatively, the structuralistic park is a construct of both the mind and the object, which is a process regarded as complex. Notably, complexity suggests that the complex reality of destroyed and fragmented urban spaces may be difficult to restore. Accordingly, the aim of German structuralistic parks is not simply to improve the ecological and social conditions from the perspective of supporting functionality. The complexity indicates the ongoing "decision-making process" (Latz 2008a) in park planning and design.

- Diversity

  The diversity and difference of space are generally understood based on the acceptance of the diversity of urban society. This point has been expounded in the third Chapter of German urban landscape analyses. The interaction between urban space and complex social construction leads to the difference. Specifically, the spatial diversity of structuralistic parks is largely embodied in multiple spatial forms and categories with distinguishing elements. At the ecological level, there is also biodiversity for ecological stabilization and dynamic balance, as well as the conservation of natural resources. Briefly, the German structuralistic park diversity manifests in aspects of society and ecology.

- Sustainability

  Three key layers—interpreted history and memory, spaces ready for social appropriation, and a current presentation of nature—usually serve as the foundation upon which sustainability is generally reflected and also provide insight into structuralistic parks. For instance, the Clear Water Canal in Landscape Park Duisburg-Nord for the New Old Emscher integrates three essential elements from the standpoint of sustainability, such as *information* layers in the landscape understanding of palimpsest, diverse social space, and the new nature integrated aesthetic and technological qualities. In order to develop nearby sites both now and in the future, structuralistic park spaces, which include those for history, perception, society, and ecology, are sustainably used. The environmentally friendly spaces develop gradually. In the transformation, the unseen, the unwanted, the leftover on post-industrial sites would come back to new life by planning and design. They would undergo a radical conversion regarding functions and meanings, and therefore achieve sustainable development through rediscovery and reuse. As Peter Latz stated, new uses and structures produce a creative tension and allow new meanings to emerge, quoted by Wolfram Höfer and Vera Vicenzotti in their article Post-industrial Landscapes: Evolving Concepts. Thus, the sustainable transformation of structuralistic parks is realized and the recycling landscape is formed.

- Appropriation

Large-scale urban park spaces are prepared for social appropriation in a variety of ways throughout daily life through planning and design (Latz 2016b). The value of structuralistic parks for the general public rests in social appropriation. The diverse and uncertain urban life is in progress. To meet the diverse desires and needs of individuals, groups, communities, and society for various uses, a variety of spaces are supplied for urban activities, programs, and events. For instance, visitors can view the F60 spoil conveyor bridge, a technical marvel from the 1990s that is now referred to as a "Horizontal Eiffel Tower". Figure 47 shows how the lighting installation by artist Hans Petr Kuhn, which was added to the site, improves the visual effects at night and gives visitors an outstanding sense of space. The interaction between space and people over time also contributes to the continuing development of the diverse spaces. In addition, the social appropriation process exhibits the virtues of autonomy, diversity, equality, and liberty.

- Identity

What has always been emphasized is the structural identity of German large-scale urban parks. It is based on the design philosophy of "decoding," understanding and re-interpretation, and "new syntax of landscape" (Weilacher 2008). The landscape of the park is a rational construct whose layers of *information* contained in its structure were preserved and transformed (Latz 2013b). The understanding of structural identity demonstrates that the derelict industrial land could be reorganized and transformed rationally based on inherent material and spatial connections, without destroying the features of a specific site.

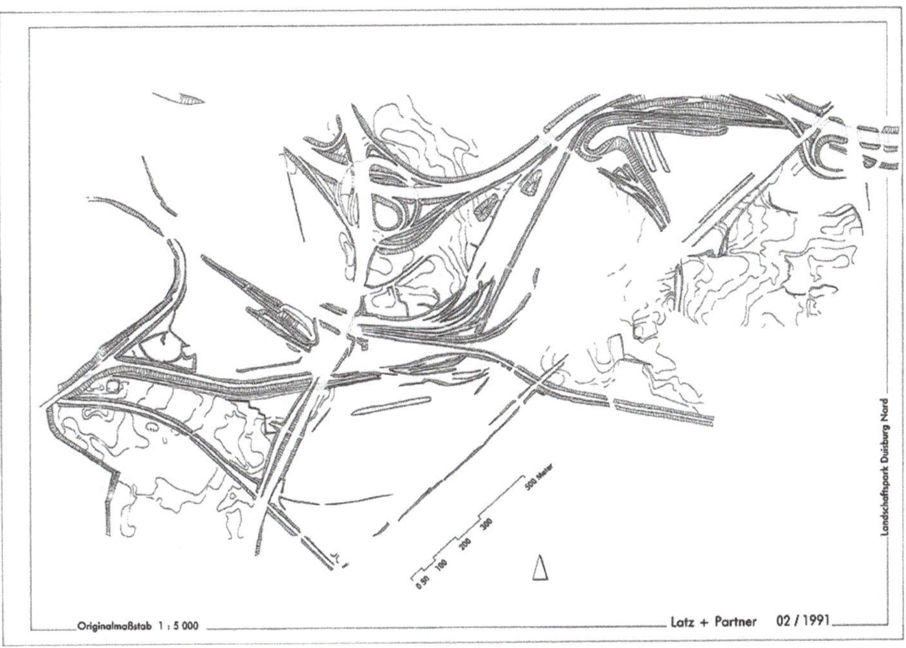


**Figure 47.** The hilly terrain of Landscape Park Duisburg Nord organized through complex site elements with the contextualistic–structuralistic approach. Source: Photo courtesy of ©Latz + Partner; used with permission.

**Figure 48.** An impressive night-time experience on abandoned F60 conveyor bridge in IBA Fürst-Pückler-Land. Source: Figure by author.

### 4.2.6. Practice: Projects of Structuralistic Parks

Two essential projects are described in relation to the German structuralistic parks. The about 230-hectare Landscape Park Duisburg-Nord in Duisburg-Meiderich (Weilacher 2008), where a complex network of industrial structures transforms into a landscape, is regarded as a prototype for German structuralistic parks. Here, a number of bullet points are presented to illustrate Peter Latz's expansive project, including serving as one of Emscher Park's projects, creating linkages with urban surroundings, using a planning and design approach, and fusing technology with natural processes.

Accordingly, the approximately 210-hectare Munich Riemer Park (1995) is an integrative spatial reconstruction project from a regional standpoint. The idea of space is further examined in terms of the "Compact", "Urban", and "Green" frameworks, as well as the open spatial structure of parks that are influenced by the morphology and components of the cultural landscape in the *urban region*. It will show how the creation of parks is integrated with various land uses to dispose of infrastructure, businesses, homes, and green open spaces in a sensible manner.

Landscape Park Duisburg-Nord

With the German urban renewal program, from 1989 to 1999, the original "rust belt" was transformed into green corridors. A regional Emscher Park as a strategy of development was presented as one of the IBA's important projects. Given the requalification of the Emscher basin and the solution to the post-industrial problems in aspects of society, ecology, and economy, a continuous spatial system was intended to be built by combining and developing disparate green spaces, and Emscher Park was formed for that purpose. The ecological restoration, re-naturalization, green corridor network building, and new housing development are thus all key themes of this project drawn up by the IBA research group (Fabris 1999). According to Sabine Auer (2010), who proposed the motto "Think Green" in the reconstruction of landscape, it serves as a "green connection" throughout the residential regions of the Ruhr valley. For instance, the extensive Emscher green connection can be perceived from one of Nordstern Park's towers in Gelsenkirchen. The various landscape elements and structures are made up of decommissioned buildings, mines, bridges, canals, terrain, and green spaces (Figure 49).

**Figure 49.** Seventeen cities interconnected through a green corridor of Emscher Park, Gelsenkirchen. Source: Photo by ©Danzi Wu 2015; used with permission.

The basic principles of development are to protect, join together, and improve existing open space to create new kinds of park at old sites while establishing the area's own park infrastructure and integrating many individual projects into a coherent park for the whole region (Rossmann 2009). These are more like devising strategies at the regional level and cannot be simply paraphrased with the vision of the park itself.

In the process of the Emscher Park project between 1989 and 1999, IBA managing director Karl Ganser highlighted that the "landscape" would be "the focal point of the urban region deliberations" (Siemer and Stottrop 2010). He added that "reconstructing landscape is by no means an isolated problem for old industrial areas. All Europe's major conurbations are happily building tomorrow's discussed industrial areas in their extensive suburban zones" (Ganser 1991). What has been firmly entrenched among German planners through this project is "the approach of using landscape as a long-term and highly effective factor of regional change" (Kolkau 2002). In short, "landscape" became the central factor of the structural transformation and development of the Ruhr district. The German understanding of large-scale parks grasped in urban regions is also demonstrated.

Against this background, the Landscape Park Duisburg-Nord planned and designed by Latz + Partner has been analyzed continuously as among the most essential projects. It stands for a completely new category of parks (Godau and Heinrich 2010). Peter Latz provided few clarifications about the site presence of Landscape Park Duisburg-Nord in the 2004 unpublished lecture "The Metamorphosis of the Twentieth Century's Landscape" at the University of California, Berkeley, quoted by Judith Stilgenbauer in 2005. He offered his opinion on structuralistic parks:

> "The park is not a park in the common sense, not easy to survey, not clearly arranged, not recognizable as a whole. According to its situation amidst chaotic agglomerations and infrastructure lines, it appears as a torn figure with numerous different aspects." (Stilgenbauer 2005)

Furthermore, the analysis and establishment of interconnections between the park and urban surroundings are regarded as the first and decisive step in Peter Latz's planning and design. The working manner has been formulated by Peter Latz:

> "When we began to work on the design task, rather than first imagining a park, we examined what would be visible from the future 'park'. [ ... ] We used an analysis plan to depict a panorama with all the elements that could be seen from the area we were working on. From the opposite direction we recorded all of the elements within the landscape park that could be seen from the outside." (Latz 2008b)

Peter Latz described a "panorama" with almost all valuable elements and connections between the interior and the exterior, as demonstrated in Figure 50. Specifically, there are "dark grey areas in the landscape park orienting to the outside", "special places" containing landmarks in red and "horizon", to which the park's externally oriented areas relate", and "streets, which relate to the blast furnace plant" (ibid.). Moreover, the remaining blank spaces in the projections are evaluated to be "quite hidden in character and not visible from the outside" (ibid.) and therefore unimportant for the further planning and design. For the interconnectivity,

Peter Latz considered the essential factors in urban regions broadly: the horizon, transportation system for convenient access, line of sight, and local distinct landscape elements that adequately represent regional landscape image and cultural characteristics.

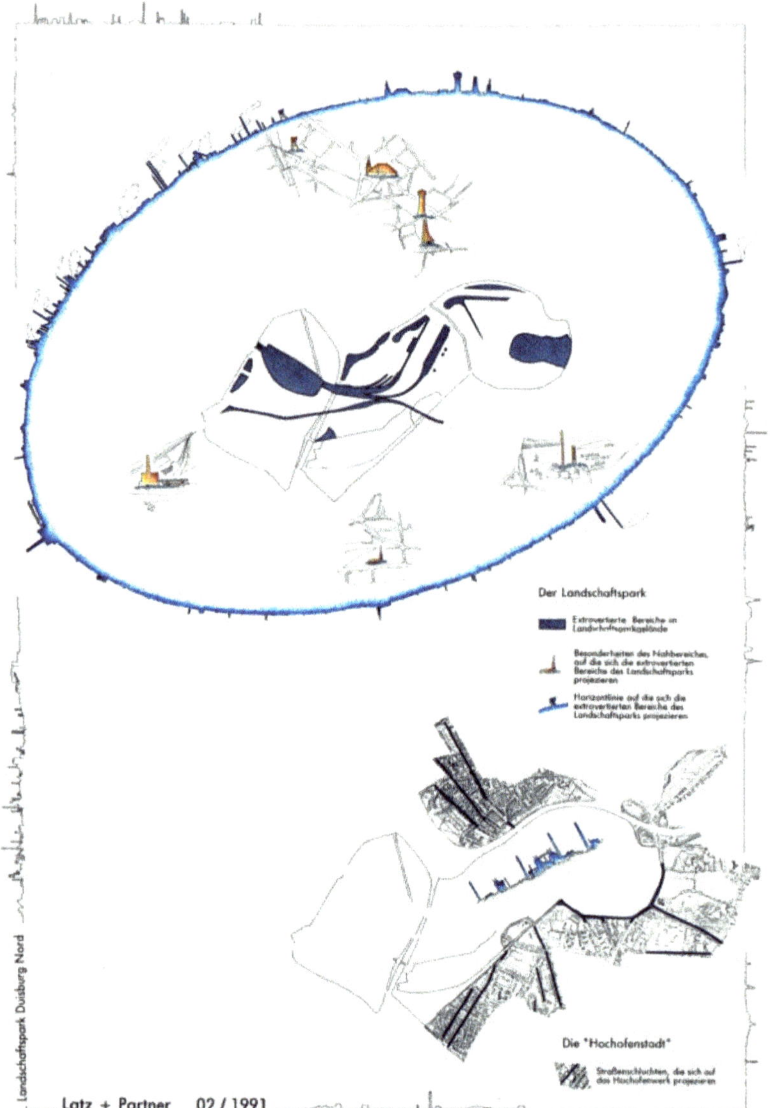


**Figure 50.** The relationships between Landscape Park Duisburg-Nord and its surroundings in the urban region. Source: Figure courtesy of ©Latz + Partner; used with permission.

Concerning the external–internal interconnections, one of the most prominent and direct ways is to build visual relations. In the park, old blast furnaces are marked clearly as both landmarks and linking elements, drawn by Peter Latz in Figure 51. The above analyses reveal that certain key nodes benefiting the interconnections are properly placed in a new park structure.

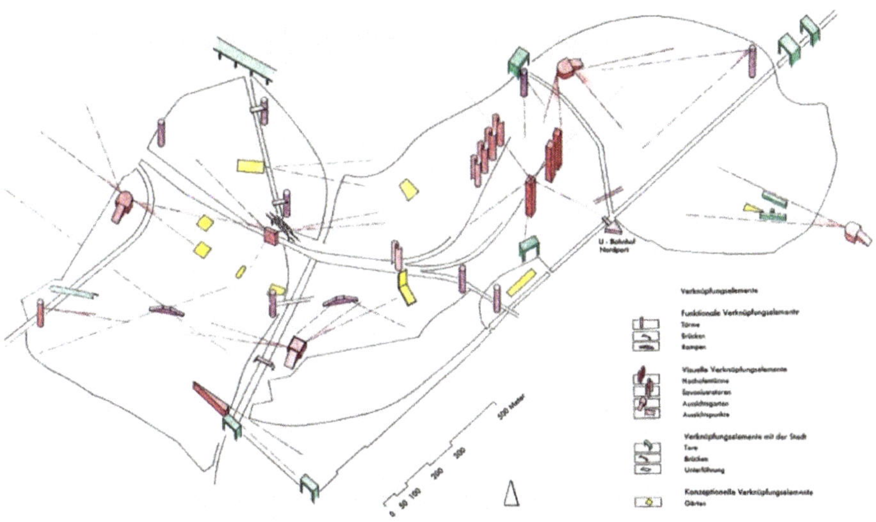


**Figure 51.** Building visual relations between Landscape Park Duisburg-Nord and its surroundings. Source: Figure courtesy of ©Latz + Partner; used with permission.

Importantly, the *structuralistic approach* is reflected in Landscape Park Duisburg-Nord. It was the first time that the design team did not work on a conventional general plan but strived to depict the park as an abstract structure and to pinpoint subspaces to be developed following certain sets of rules: the railway park, the water park, the city promenades, and so on. Importantly, a mesh of industrial structures became the landscape. In the work *Rust Red*, Peter Latz even remarked: "we never wanted to draw an overall plan for Duisburg-Nord as the medium only shows one layer. As feared, the plan depicts the actual chaos. Order cannot come from chaos, it can only be understood in abstraction" (Latz 2016b). This expression not only suggests the importance of the *structuralistic approach*, but also corroborates the idea of critiquing the image in the conception of a contemporary new park.

The categories as structural elements are "linked together visually, functionally, through ideas or symbolically, using the smallest possible interventions, special connecting elements, ramps, steps, terraces or gardens" (Weilacher 2008). Additionally, a complex network of industrial structures becomes the distinct cultural landscape for the future based on the "overlay and connection of independent conceptual layers and structural elements" (ibid.), as shown in Figure 52.

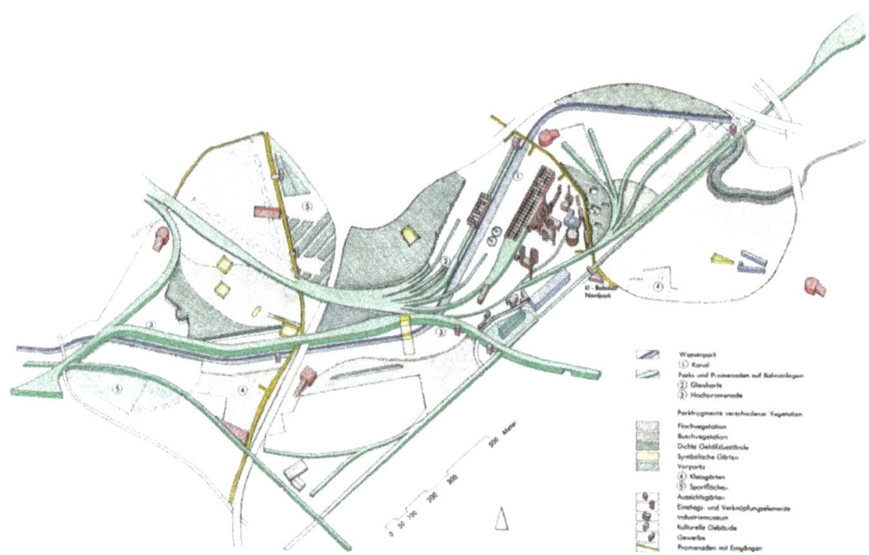

**Figure 52.** The abstract structure of Landscape Park Duisburg-Nord developed through overlay and connection of layers and structural elements. Source: Figure courtesy of ©Latz + Partner; used with permission.

Finally, the landscape project displays Peter Latz's understanding of nature and its processes, as well as the relationship between nature and technology. With the dynamic processes of ecosystems revealed by increasing ecologists, "Latz + Partner's conception of natural systems reflects the shift" (Rosenberg 2007). It is the shift from equilibrium to nonequilibrium paradigms articulated in the part of North American organic parks.

With "artefacts as a symbol of nature", the natural processes supported by the technology are expected to be cultivated in the structuralistic parks. It is similar to Latz's explanation of the "water canal" in this case. The old Emscher wastewater canal becomes the water park and water promenades, after the ecological process of restoration in Figure 53. The water canal is an "artefact aiming to introduce natural processes in a devastated and perverted situation", as depicted in Figure 54. These processes operate according to the rules of ecology but are initiated and sustained by technological means. Man uses this artefact as a symbol of nature with the feature of wilderness while being responsible for the process. "It is the most natural and the most artificial system at the same time", quoted by Udo Weilacher in *Syntax of Landscape*. In the cultivation of natural processes, the German structuralistic parks are regarded as eco-machines for shaping self-organizing and resilient urban nature, described in the similarities of two large-scale urban parks in Chapter 5.

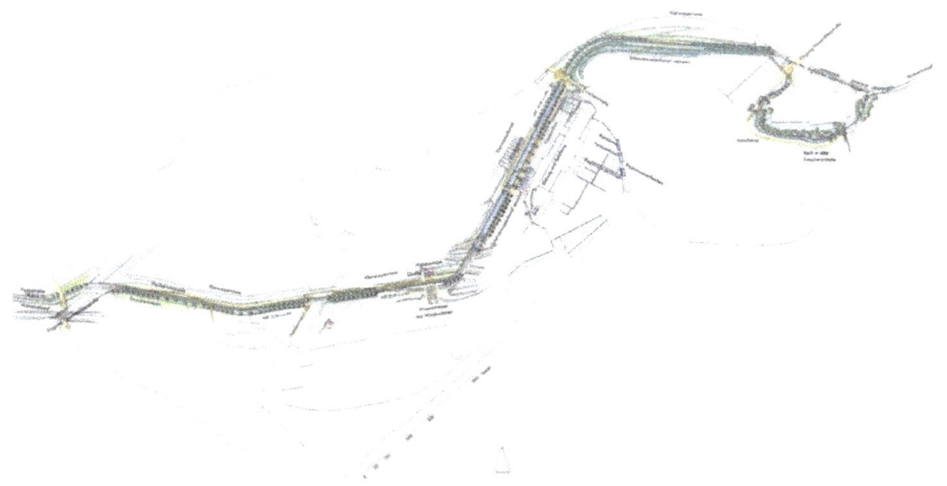

**Figure 53.** The "water canal" designed for ecological restoration in Landscape Park Duisburg-Nord. Source: Figure courtesy of ©Latz + Partner; used with permission.

**Figure 54.** The natural processes supported by the ecological technology are shaped over time in Landscape Park Duisburg-Nord. Source: Photo by author.

Concerning the inextricable relationship between nature and technology, Peter Latz once regarded nature in harmony with technology in the book *Syntax of Landscape*:

> "So technology and nature not as a contrasting pair, as in early Modernism, but technology and nature in accord. Here I am interested in a possible congruence within the ecological concept. This is nothing to do with the need for harmony; no, the technical idea is to try to integrate nature sequences as much as possible, and to let nature be nature. On the other hand, nature we create artificially must allow us to find an aesthetic language that is identical with the technical one. [ . . . ] I am absolutely allergic to the idea

that nature should reconquer something for itself. [ ... ] We have to keep a hold on technology, and integrate it into our environment." (Weilacher 2008)

On reclaimed ground, where natural processes predominate the dynamic development, following the termination of continuous, intense human control or influence, it is clear that *successional nature* (German: *Sukzessionsnatur*) generated through the technical approach frequently occurs. The processes are a reflection of German urban ecology and landscape architecture returning to nature. The abandoned site's original, authentic character and ambience are consistent with the wilderness character of *successional nature*.

Latz + Partner's succession of the park is envisioned at various stages in terms of "unmanaged free succession", "managed succession to halt current stages or recreate past stages" and "succession controlled by usage", as well as at multiple levels, such as gardens regarded as a symbol of transformation, green spaces with trees or shaped hedges, and paths and large areas of gravel turf (Latz 2016b). In the design of vegetation, Figure 55 shows various vegetation types, including the type without intervention for keeping the wilderness with vigorous plants, horticultural planting in selected small areas using trees in rows or on a grid, aquatic planting along the banks, and the rotation of crops on agricultural land. According to Latz + Partner's idea of "vegetation patches", the vegetation landscape plan is created by controlled and uncontrolled succession, as illustrated in Figure 56.

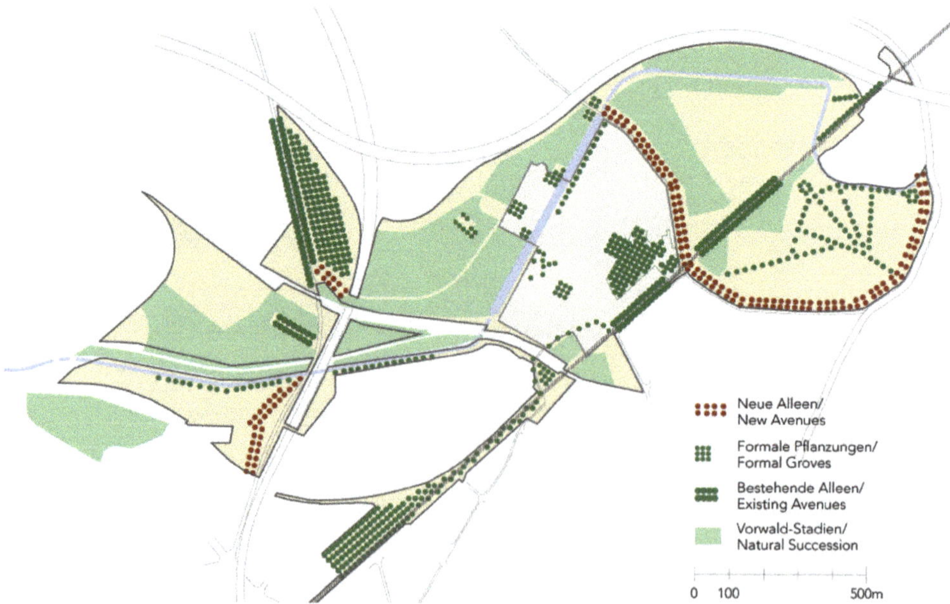


**Figure 55.** The various vegetation types of Landscape Park Duisburg-Nord. Source: Figure courtesy of ©Latz + Partner; used with permission.

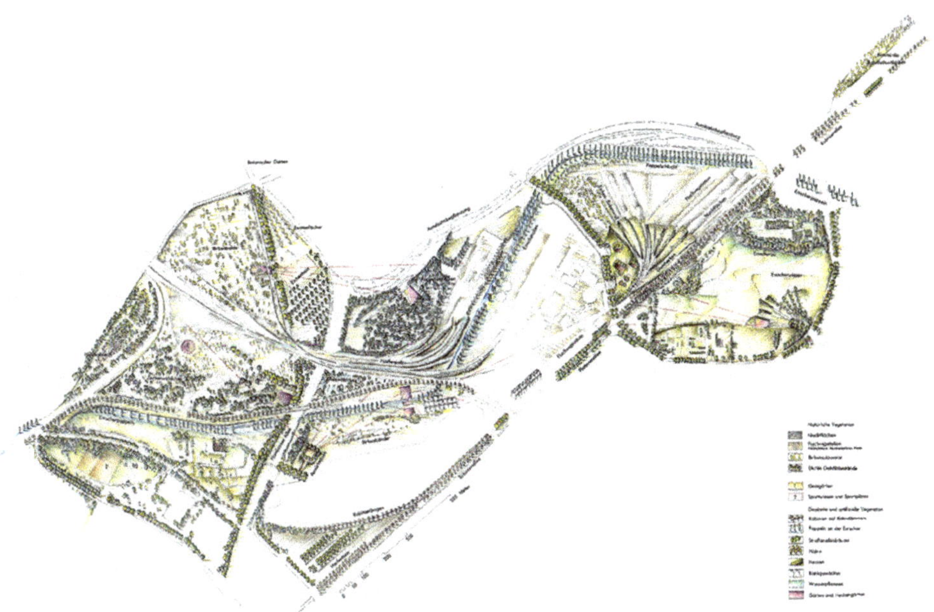

**Figure 56.** The vegetation landscape plan of Landscape Park Duisburg-Nord. Source: Figure courtesy of ©Latz + Partner; used with permission.

Riemer Park

Munich-Riem Airport was housed there until 1992, and Munich-Riemer Park is the second project. In the German urban landscape idea, it might represent not only the "socio-spatial features of the landscape" but also the landscape's structure as a provider of structures on the outskirts of cities (Schöbel 2015). Munich should give the German landscape architect Fredrich Ludwig von Sckell credit for creating the two substantial urban parks English Garden and Nymphenburg Palace Park. They are now beloved classical large-scale parks with the idea of imitating ideal nature and bringing nature into the city (Figure 57). According to von Sckell (1825), in order for gardens or parks to function well, they must be in proportion to and employ the same formal language as their natural surroundings. Consequently, in the typical German landscape park, we notice a pleasing landscape made up of curved paths, clusters of trees, large lawns, mild hills, and lakes. Particularly, in the construction of contemporary new urban large parks, such as Riemer Park (Figure 58), the straight routes, huge gravel plains that extend to the horizon, clusters of horizontal planting, and clearly defined agricultural and forestry land, however, result in the creation of a new and potent landscape structure in the north-eastern part of Munich.

**Figure 57.** English Garden and Nymphenburg Palace Park in Munich, Fredrich Ludwig von Sckell, 1789 and 1823. Source: Photo by author.

**Figure 58.** Riemer Park showing a new structural language in Munich, Gilles Vexlard, 1995. Source: Photo by author.

The winning proposal, the Riemer Park master plan, comes from Paris Latitude Nord led by French landscape architect Gilles Vexlard. Through the 1995 international park design competition, the derelict 560-hectare Munich Airport land was turned into a new and modern city district of Munich, called Messestadt Riem. The urban project for site transformation could be materialized through blending business, residence, trade fair, infrastructure, and green open space (LMRSB 1998). Moreover, it effectively motivates the Munich urban development toward the east in a sustainable way. Riemer Park is not merely a large-scale urban park project. It is "one of the City of Munich's biggest current urban development projects" (Zöch and Loschwitz 2005).

The spatial conception of Messestadt Riem is set up based on an essential principle of the whole area disposition. It is defined as "Drittellösung" in German, which indicates a three-part solution (LMRSB 1995). According to land use, it is explained as "one-third of the area zoned residential, one-third allocated for industry or business development, and one-third for parks and open space" (Schegk and Wilk 2007). The unique solution fully conforms to the leitmotif of Munich sustainable urban development, "Compact—Urban—Green" (LMRSB 2005), following the concept of "Munich Perspective" by the Munich City Council in 2008. Hence, Lutz Hoffmann explained it in the essay 850 Years of Urban Development in Munich:

- Compact: the use of "urban space sparingly by compactly and densely."
- Urban: "a lively mix of residence, worksites, shopping and recreational venues."
- Green: "an attractive array of open spaces and green areas to improve the natural environment and the recreational potential." (Hoffmann 2008)

For instance, Riemer Park and the settlement are connected by the green open space, which is situated in the center north–south green corridor with rough grassland and woody structures at the outskirts. The large park is connected to urbanity by a green connection. According to the ecological concept, the new settlement follows a climatic model of ventilation and fresh air delivery, as shown in Figure 59.

**Figure 59.** Following the leitmotif of "Compact—Urban—Green", the green corridor links the settlement to Riemer Park. Source: Photo by author.

Buildings and green open space are constructed for the Riemer Park project at diverse scales, such as park, plaza, courtyard, and garden, which are strongly mixed for a variety of urban functions and social uses. In the western half of Riemer Park, for instance, a number of sunken gardens, such as the well-known *Cell Gardens* (German: *Zellengärten*) and *Leaf Garden* (German: *Blattgarten*), are developed near to the settlement and serve as spectacular locations in the idea of *Change of Perspectives* for creative perceptions of urban nature during BUGA05, as shown in Figure 60. This is according to the functional principle of short ways, which is followed by blending with a reasonable density in Figure 61. It also illustrates a fundamental understanding of Henri Lefèbvre's idea of the *social production of space*, illuminating how social space is created.

**Figure 60.** A series of sunken gardens for diverse perceptions and uses of spaces in Riemer Park. Source: Photo by author.

**Figure 61.** The dwellers can easily access Riemer park's extensive open spaces and Badesee lake via straight pathways and boulevards. Source: Photo by author (right) and ©Danzi Wu 2016 (left); used with permission.

From this perspective, differential social relationships are reflected in the organization of diverse spaces. A graded open space system (German: Ein abgestuftes Freiraumsystem) is generally set up and developed (LMRSB 1998). The system contains diverse spatial forms for two types of social organization: community (German: Gemeinschaft) and society (German: Gesellschaft), from the core area named Willy-Brandt-Platz, located in the south of the east–west axis Willy-Brandt-Allee, to the southern open park. They are classified as private gardens for individuals and families; green areas between buildings serving local neighborhoods or groups; plaza, such as Menschenrechte for public life; Willy-Brandt-Platz for German urban society's mixed uses; and the park as a green open space, intensively connected with residential areas, available for urban residents and visitors (Figure 62).

**Figure 62.** The environmentally friendly neighborhood space and public space for mixed uses in Riemer Park. Source: Photo by author (right) and ©Danzi Wu 2016 (left); used with permission.

Furthermore, the park's spatial structure is considered based on the principle of "Drittellösung". According to the given assignment during the 1995 Messestadt Riem München Competition, the first thing for the construction of the park alone is to develop a landscape structure as its high identity, so as to meet ecological requirements in terms of the climate and biotope network, provide an open space for the 41,000 residents, and integrate the fenced area of the old airport into the existing system of green corridors (LMRSB 1995) (Figure 63).

**Figure 63.** A wealth of flora and fauna habitats and green corridors are shaped in the unique network of Riemer Park. Source: Photo by ©Danzi Wu 2016; used with permission.

From one perspective, the park formulates the transition between large-scale landscapes of forests and wilderness and reconstructs the historical meadow structures (Schöbel 2015). The land structure of the park embraces the topography and local characteristics, specifically represented as Munich Gravel Plain during the glacial period, parcels of land naturally divided into cultivated and woody land, massive woods made up of oak–pine, and oak–hornbeam in this region. Combining these on-site landscape elements, Gilles Vexlard added that "the power of the Munich landscape is the distance" (Schegk and Wilk 2007). The distance indicates a perceived openness in the landscape. Accordingly, he sketched the conceptual park structure in extensive linear forms to symbolize the Munich cultural landscape, "agriculturally imprinted cultivated landscape of Munich's east" (Zöch and Loschwitz 2005) (Figure 64).

**Figure 64.** Riemer Park, open to the surroundings with characteristics of Munich cultural landscape. Source: Photo by ©Danzi Wu 2013; used with permission.

The park is fully open to its environment and links the overall Riem area to surrounding villages and Munich cultural landscape through a network of inclined routes, strips, and bands of native woods and shrubs extending to the horizon. This scene is generally described as "a park without borders" (German: "ein Park ohne Grenzen") (LMRSB 2009). In conclusion, the open structure reflects the Munich landscape image, forms a dialog with the surroundings, and offers citizens the perception of openness and freedom.

From another perspective, the open structure considers airflows and assumes a certain ecological function. It takes the prevailing wind directions throughout the whole site into consideration. At least a 400-meter-wide fresh air corridor of the park has an effect. Plenty of fresh air would be supplied through the air corridor from the Ebersberger forest situated in the east towards the Munich City under the lack of air exchange in the weather condition (ibid.). The open structure of Riemer Park takes on spatial qualities of openness, freedom, and identity, as demonstrated in Figure 65.

**Figure 65.** Along the lake (Badesee), the strips and bands of native woods emphasizing the spatial structure of Riemer park. Source: Photo by ©Danzi Wu 2017; used with permission.

Another example of expansive urban parks is the German design paradigm of structuralistic parks, which is presented in this chapter. The German parks in the site transformation are strongly tied to former industrial sites with underlying and collected *information* that needs to be decoded, understood, and handled, as well as their surroundings and even urban regions in the large thinking. They take a comprehensive approach to the entire region as well as just the one abandoned site.

Peter Latz, meanwhile, gives his own critical interpretations of architectural *structuralism* with regard to its significance in the Netherlands in particular and the *minimal intervention* by Bernard Lassus and Lucius Burckhardt adopting the contextualistic–*structuralistic* approach. The unique *syntaxes*—abstract structures that overlay and connect independent conceptual layers and structural elements (Weilacher 2008)—by which diverse, free social spaces in everyday life would develop over time are critically examined and shaped by the urban landscape in large-scale parks in a new and careful way.

The German contextualization of Peter Latz in contrast to the North American *cultural imagination* of James Corner, to sum up, is one of the most crucial distinctions reflected in two large-scale urban park paradigms. Their connections go further than this and are more clearly shown in two areas of similarity and difference. Therefore, the next Chapter's cross-cultural comparison will examine the relationships between two large-scale urban park paradigms from North America and Germany.

## 4.3. Shan-Shui Parks in China

Ultimately, the discussion around the two above-mentioned large-scale park models plays a role in promoting the rethinking of Chinese shan-shui parks. In Chapter 2, it was demonstrated that shan-shui parks arise from the traditional conception of the shan-shui city and the idealized urban landscape. From the perspective of the design approach, shan-shui park as a typical model has been met with a wide range of applications, such as country parks, forest parks, and heritage parks, etc. Chinese landscape architects are accustomed to basing the generic shan-shui spatial pattern on their individual understanding of shan-shui as the cultural essence of landscape architecture as an independent discipline. For instance, Figure 66 shows the shan-shui landscape images of Bei Wu Country Park and Shougang Industrial Heritage Park built in Beijing City. They are common in featuring picturesque mountains and traditional buildings through multiple spatial visual axes. Regarded as one of the Chinese gardening techniques of "borrowed scenery" ("借景" in Chinese), it incorporates the elements and views derived from landscape surroundings. Meanwhile, the spatial configuration of shan-shui sticks with the principle of "North Mountains and Southern Waters".

**Figure 66.** The static landscape images of shan-shui parks, Beijing. Source: Photo by author (right) and ©Jiani Li 2021 (left); used with permission.

However, when it comes to the question about the modernity of Chinese landscape architecture as discussed in Chapter 3, what are the modernity of shan-shui parks and their approach to planning and design like? Can the traditional park design paradigm of this type be better applied to the twenty-first-century parks built on post-industrial sites to resolve the socio-ecological issues in the ever-changing urban landscapes? These questions lead us to reconsider shan-shui parks through critical thinking. Furthermore, for Chinese large-scale urban parks, it remains necessary to experiment with various park styles gradually and explore the local design languages in a context of cultural clash.

Regarding the application of shan-shui parks, the urban sprawl and ecological environmental crises unfolding in most Chinese metropolises has led to the initial systematical planning in the form of country parks, by drawing on the constructive experience of Hong Kong. As one of the earliest cities to embrace country park planning and design in suburban areas, Beijing is almost successful in meeting the demand of urban residents for recreation, ecology, history, and forests. Since the 2000s, Beijing has been vigorously constructing country parks under the country park development program as a way to implement the greenbelt strategy. This is aimed at enabling ecological services, maintaining the well-developed urban spatial structure, achieving the coordinated development of urban and rural areas, and restricting unordered urban sprawl.

In recent years, the emergence of abandoned industrial sites has prompted the movement of urban renewal across numerous Chinese cities to reduce the incremental construction within urban space. In this process, Chinese landscape architects begin to apply the abstract shan-shui spatial relationships and exercise some specific elements to demonstrate the modernity of the shan-shui landscape, as well as to deal with the renewed planning and design of abandoned industrial sites. In other words, the opportunity for urban renewal may lead to the exploration and redevelopment of shan-shui parks with the distinctive design ideas and approaches of Chinese landscape architects.

### 4.3.1. Vision: Parks in Urban Fringes

Most importantly, under the strong influence of shan-shui culture, the concept of Chinese country parks emerges in Hong Kong as one of the large-scale shan-shui park models. Due to British colonial rule, Hong Kong followed the term in the United Kingdom, with most of the country parks built during the 1970s recognized by its government, in accordance with the Countryside Act 1968.

In Hong Kong, there has been some progress made by AFCD since 1976 in terms of the designation, development, and management of country parks in line with the Hong Kong Country Parks Ordinance. Under this framework, a large-scale park system consisting of 24 country parks was put in place in 2013. Within the extensive park system, the country parks are planned and constructed to consolidate and preserve those essential landscape elements and natural resources, including hills, woodlands, wetlands, islands, reservoirs, and coastlines, as illustrated in Figure 67. In most cases, the country parks built in Hong Kong are characterized by the conscious choice of various urban areas reliant on the advantages of location, topography, natural and ecological resources, and urban infrastructures.

The county parks in Hong Kong set a classic example for the planning and design of contemporary large-scale parks in the urban–rural fringe, and even the wider urban areas. The country parks are capable of offering protection to the vegetation and wildlife, preserving and maintaining the buildings and sites of historic or cultural significance, and providing the facilities and services that the public need for enjoyment. It is quick for the Hong Kong version of country parks to be recognised as a unique large-scale park model by other cities, due to its remarkable success in integrating natural resource conservation with urban recreational activities in many parts of the city. This park model contributes to the practice of utilising and managing the urban ecological environment. Meanwhile, urban landscapes are better shaped in the overall urban region through regional morphology as well as natural landscape elements and their characteristics.

**Figure 67.** Country parks in Hong Kong manifesting the ideal settlement of shanshui city. Source: Photo by ©Danzi Wu 2018; used with permission.

By drawing on the experience gained by Hong Kong, Beijing City started, from 2007, to develop its own country park system as part of the 2004–2020 Beijing master plan and green system plan. In particular, country parks have routinely been regarded since 1958 as the essential components of planned greenbelts according to the proposed greenbelt strategy. This leads to the emergence of a series of country parks at multiple scales and their distribution in both inner and outer greenbelts.

The greenbelt strategy adopted by Beijing City was proposed in the 1958 Regulating Plan to put the concept of inner greenbelt into action. As suggested by a planned greenbelt, a broad zone dedicated to green open space should be created to restrict urban sprawl, prevent excessive development, coordinate the development between urban and rural areas, and improve the urban ecological environment.

However, there is no end put to the practice of urban development encroaching on the planned greenbelt with the notable reduction in the inner greenbelt from 314 km$^2$ in 1958 to 240 km$^2$ in 1992 (Li et al. 2005a). This is attributable to unplanned urban sprawl. It is usually driven by unrelated governmental or politically motivated actions and by private economic activities and speculation (Stokman et al. 2008). In the meantime, "there are not enough resources to devise and implement regulatory policies and tools to control the pace of development" (ibid.).

Nevertheless, to further organize green open spaces, the second-stage greenbelts, with more country parks, continued to be planned in 2004. Currently, there are two-layered greenbelts or two greening isolated areas. One is the inner greenbelt situated between the fourth and fifth ring roads. It is located at the transition area between the inner city and the surrounding satellite towns. The other is the outer greenbelt situated between the fifth and sixth ring roads. It is located in the transition area between the urban and the rural areas (Li et al. 2005a).

A total of 25 country parks distributed in the inner greenbelt existed up until 2011. In 2022, the number of them rose sharply to 81. However, most of these country parks are less interconnected at different scales, as demonstrated by the scattered distribution of major country parks in the present inner greenbelt. The vast majority of them cover less than 100 hectares, with the 680-hectare Olympia Forest Park as the largest one. In addition to these parks, there are another four country parks at a regional scale that were planned in the extensive area of the outer greenbelt in 2007. Currently, there are 40 country parks distributed in the outer greenbelt, covering 160 hectares on average.

Notably, the existing Beijing country park concept shows a far less significant relationship with the shaping of urban landscapes with regional cultural characteristics. Therefore, further exploration is required for their shan-shui structures and features at both local and regional levels. Thus, this study aims to study this shan-shui park model as a contemporary form of urban landscape, by critically referencing the other two large-scale park models from developed areas and by exploring their own specific park identities and design approaches. Like the cases of North America and Germany, it is possible for the above challenges and tasks arising from the shift from industrial to post-industrial society to move Chinese shan-shui parks forward while promoting the development of urban landscapes both in theory and practice through the critical approach.

### 4.3.2. Reflection: Parks in Urban Landscapes

As for Chinese large-scale urban parks, they should be reflected in the understanding of contemporary urban landscapes. There is a significant urban–rural separation shown by the spatial structure in the megalopolis, despite the policy proposed by the Chinese government in 1956 to foster a "New Socialist Countryside". The policy was aimed at a radical improvement in terms of economy, infrastructures, culture, and environment. The understanding of urban landscapes among Chinese landscape architects is subject to the profound influence of the current state of the urban spatial structure.

Landscape is what clearly reflects the binary structural opposition in practice. Regarding the single effect of city beautification, it is through the highly artificial and ornamental approach that the landscape is organized and valued in most Chinese cities. Differently, the extensive countryside is where the chaotic and underestimated rural landscape expands. Objectively, the root cause of differential landscapes is the considerable gaps in social structure, economic development, and urban infrastructure. In Chinese landscape architecture, the two opposite landscapes share the theoretical foundation, that is, city beautiful. Originating from the global City Beautiful Movement in the 1893 Chicago's World Columbian Exposition, the concept was put forward by the American journalist and urban planning theorist Charles Mulford Robinson in 1903 (Yu 2012). Transferred into Chinese landscape architecture, city beautiful has been acknowledged as one of the most imperative ends for urban landscape planning and construction.

With the changes in urban structure and society, however, it is also necessary to expand the understanding of urban landscape certainly. The boundary between the urban and the rural may likely end up being dissolved as time goes forward. What the future holds may be similar to the urban phenomenon seen in two developed regions as explained above. At this stage, the top priority in the gradual transition of the Chinese city and society is to break the barriers between the urban and the rural, and those between the city and the landscape, by ditching the dualistic thinking pattern, and to further improve the reconciliation between them. That is to say, the strict spatial and conceptual division should be abandoned to promote interconnections.

Meanwhile, it is inappropriate to simply regard the concept of Chinese urban landscape as an urban landscape with a rising level of urbanization. To be specific, it is the landscape limited to being in and around the city center where a large number of tasks were performed by most landscape architects who undertook plenty of work. By contrast, the countryside landscape could not be ignored in planning and design. Thus, it is necessary to expand the concept of urban landscape if the city is considered as a unified and interconnected system. Moreover, the landscape is expected to fulfill its function in a broad sense. Since 2021, territorial spatial planning in China has begun to experiment with the syncretic integration of spatial resources based on urban landscape construction and reorganization of the blue–green spatial system at different levels. In this process, landscape architects need to be involved in a wider range of tasks.

Therefore, it is a necessity that a critical approach is taken to view shan-shui parks from a qualitative perspective. Similarly, the above five qualities as reflected in both North American and German park design paradigms, including complexity, diversity, sustainability, appropriation, and identity, are reconsidered as follows.

- Complexity

  As described in the previous chapters, the complexity of the North American organic park arises from the established complex systems. It is identified as a dynamic, messy, contaminated park site. Given the close link between information and elements on the park site, the complexity of the German structuralistic park is manifested in the highly complex design process, in combination with existence and invention. As suggested by two pathways in developed regions, the planning and design of large-scale urban parks are driven by a complex practical environment as analyzed within their own landscape architecture.

Nevertheless, the concept of city beautiful and the traditional view of nature decrease the attention paid to the complexity of shan-shui park. On the contrary, they are often conceived in an ideal condition of urban society and ecology. This leads to a routine that the planning and design of shan-shui parks or country parks within greenbelts are dealt with in the same way as those urban parks in the inner city, which ignores the intricate man–land relations in the urban–rural fringe, and contaminated, derelict sites. When the shared challenges of urban nature and society on post-industrial sites are coped with by North American and German landscape architecture, the simplistic consideration given to shan-shui parks without the complexity is clearly insufficient for their further conception. Although the shan-shui parks place no requirement on the continuous generation of beautified landscapes in contemporary Chinese cities, sustainable urban landscapes remain essential for improving the urban ecological environment during the gradual process of social transition. In this sense, the understanding of shan-shui parks can be improved from the ecological perspective through the theoretical analysis of complexity pertaining to the organic North American model. Especially, the dynamic and process-oriented feature may trigger a radical change to the long-standing Chinese park concept in a static way and the construction accomplished overnight, without the possibility of further adaptation.

- Diversity

As revealed by the heterogeneity of North American organic parks where biodiversity is the focal point, it is requisite to maintain its largeness and spatial connections and interactions. Regarding the current shan-shui parks, the enhancement of ecosystem functions hinges on biodiversity in the face of the havoc of the urban ecological environment, and this is closely related to the way of planning and construction. Hence, their diversity can be indicated from the landscape-ecological perspective.

In essence, diversity necessitates the integration of various landscape elements into shan-shui parks. That is to say, the interconnected landscape systems are where they should be positioned. Locally, the organization of landscape elements surrounding shan-shui parks is achieved by building regional corridors between ecosystems. For instance, the spatial organization of a series of country parks in the inner greenbelt of Beijing city has no connection with the elements of their surrounding landscape, such as mixed coniferous and broad-leaved forest, grassland, water body, wetland, and agriculture. As a result, most country parks in Beijing are scattered and disconnected from the wider ecological context.

- Sustainability

In general, sustainability is supposedly reflected more in shan-shui parks for theoretical and practical exploration into the Chinese ecological ideas related to landscape architecture. Profoundly influenced by city beautiful, most of them are designed as artifacts and built with a significant ornamental feature, which results in not only the substantial consumption of human and material resources but also the higher costs of maintenance and construction.

With regard to the cognition of sustainability, certain helpful information and clues can be provided by both the North American resilience and the German sustainable utilization of remnants and coherent development of spatial quality. Therefore, it is essential to deepen the Chinese cultural understanding of sustainability for country park conception, especially from the landscape-ecological perspective.

- Appropriation

Allowing for the social appropriation taken into account in North America and Germany, both of them are beneficial in terms of daily urban life. It is through a form of self-organization in the growth and transformation of parks over time that the appropriation is achieved for the North American organic parks. In the process, the programmatic indeterminacy plays a vital role. For the German structuralistic parks, diverse spatial organization is premised on the diverse appropriation at different levels of social organization, which involves ordinary individuals, groups, communities, and society. Both focal points of social appropriation are characterized by openness, liberty, autonomy, and equality. They have a promoting effect on the rethinking of Chinese shan-shui parks.

It is from everyday life that the social appropriation arises. For shan-shui parks, the reduced appropriation may be largely attributed to the lack of accessibility and availability. With respect to accessibility, there is less connection between large-scale urban parks and infrastructure, villages, residences, or business areas. Not planned within an integrated urban system, shan-shui parks are difficult to provide an everyday social space for diverse uses. Rather than being made available, most shan-shui parks are designed as scenic spaces that feature visual landscape elements and focuse on aesthetic and sensuous qualities. For this reason, in the future, it is worth carefully considering abundant and attractive activities, programs, and events, as well as the spontaneous or organized.

- Identity

The previous discussion around the identities of two park models is based on the confirmation of them as an urban landscape. The focus of these two large park models under analysis is not placed on the concept of the park. Instead, they are similarly considered as an open, extensive, and connected landscape at the urban level with the incorporation of urban spatial structure and the expansion of urbanity into the landscape. By comparison, Chinese shan-shui parks are considered only as parks whether in theory or in practice, which is determined by the urban–rural binary structure as the mainstream urban organizational form. In this sense, the exploration of the shan-shui park identity makes it necessary to get rid of the binary thinking pattern in the practice of urban planning and design at first.

In comparison with the North American organic and German structural identities, shan-shui parks require the exploration and establishment of their own one. Aside from the concept of the park, there is difficulty in making an immediate description of their exact identity in the context of Chinese culture. It is thus crucial for them to find their own cultural identities in a critical and intellectual way, which is a suggestion made according to the experiences of developed regions. To find its cultural identity, Chinese professionals still have a long way to go.

In addition to embracing a theoretical analysis from the qualitative perspective, the rethinking of shan-shui parks about the two park models also involves a practical park case analysis conducted in Beijing City, where a country park was planned and designed by taking the opportunity of Olympic Games and an industrial heritage park was constructed by relocating the Shougang Group, as shown in Figure 68.

**Figure 68.** The remaining structures and buildings after the relocation of Shougang Plant, Beijing. Source: Photo courtesy of ©Tsinghua Urban Planning and Design Institute, Zhu Yufan Studio; used with permission.

4.3.3. Transformation: Parks for Urban Regeneration

Since the second millennium, China has experienced structural reform carried out in those post-industrial areas. China is widely known to have engaged in the international industrial market since the late 1980s, showing an impetus that has accelerated and condensed what had occurred in decades in other countries of the world in just years. From construction to their operation, industrial plants go through a life cycle that has been significantly reduced from decades into several years. Meanwhile, scientific research and technological innovation trigger the emergence of new plants, even before the demolition or renovation of the existing ones. Given the increasing significance attached to the principles of ecological preservation and sustainability, as well as the release of ecological awareness policies, a whole series of principles have been implemented in the great Asian country, which allows planners to reshape the industry and redesign the territory through multi-scalar interventions that have better recovered the models proposed in Europe and North America.

Landscape architecture is an attempt made to assist the structural and recovery modifications that can be made in cities and their compromised sectors. As a result, the large-scale park is more of a conceptual strategy intended to achieve enormous changes, which is seemingly consistent with the global trend seen in the West. Among them, Beijing Shougang Industrial Park is representative of this structural transformation. In this period, the particularity of postindustrial landscape transformation is demonstrated. Firstly, under the ecological civilization system, green space policies play a role in increasing the scale of postindustrial area renewal and accelerating its pace. Secondly, the "landscape" started to become a structural tool used to promote the development and renewal of post-industrial areas. Thirdly, the derelict areas are promoted by the top-down driving forces of ecological and environmental protection as a place of trial for post-industrial landscape regeneration.

In this circumstance, the following four bullet points are supposed to be relevant to the large-scale urban parks as a planning concept for the conversion of post-industrial sites:

- The park landscape practice provides a critical means to test post-industrial landscape theories, as well as for China to learn valuable experiences from international practices to create more iconic cases. Additionally, it presents the best opportunity to put cultural self-confidence on display and shape regional landscape characteristics.
- At the social and academic levels, the aesthetics of industrial ruins have been recognized by most landscape architects through theoretical research. In spite of this, it remains necessary to analyze and condense the concept of *urban-industrial nature* with Chinese characteristics. From a social point of view, the aesthetic turn is also anticipated to take place promptly in China. From the perspective of legislation, China should further standardize the design and construction of derelict areas by formulating more laws and policies in close relation to the renewal of post-industrial areas.
- In the course of Chinese ecological civilization construction, there are obvious system advantages shown by top-down ecological ideas. Therefore, landscape architects should pay more attention to exploring the significant ecological connotations. The regeneration of post-industrial areas will definitely contribute to the building of "Beautiful China". Chinese landscape architects should take this opportunity to elevate the critical theories of urban landscapes and parks.
- Through large-scale parks, the futuristic change in the post-industrial landscape requires an inclusive and integrated innovative pattern of thinking. An open and diversified treatment method could be provided by different disciplines and multiple environmental design elements for the transformation of those derelict areas. The key is to develop specific solutions for regionalism through "landscape".

### 4.3.4. Practice: Projects of Shan-Shui Parks

In this section, the two essential shan-shui park projects undertaken in Beijing are described in detail. An attempt made by Chinese landscape architects was to carry out the planning and construction of Beijing Olympic Forest Park by incorporating green ecological ideas, international advanced ecological techniques, and Chinese traditional shan-shui cultural elements. This large-scale urban park is regarded not only as a typical example for the construction of a shan-shui city but also a practice for three levels of the shan-shui city: fengjing shanshui, ecological shanhui, and humanistic shan-shui.

Moreover, Shougang Industrial Heritage Park is regarded as among the most successful large-scale park project undertaken on those abandoned industrial sites in China. As a prototype, it plays a role in promoting the redevelopment of landscape architecture as a profession as well as in the conservation and reuse of industrial heritage in terms of architecture. In the process of urban renewal, there are more post-industrial sites in China having attracted attention from the authorities and academic circles. The commonality is that the Olympic Games, as an important sporting event, has been taken advantage of to build large-scale parks for permanently retaining green open space in the city and sustaining the development of urban space.

Beijing Olympic Forest Park

Olympic Forest Park (680 hectares) stretching across the fifth North Ring Road is today considered one of the largest shan-shui parks in Beijing city. The whole park is divided into northern and southern halves by the ring road. The USA Sasaki Associates won the 2001 international competition for conceptual planning and design of Olympic Green composed of Olympic Forest Park, a central area, and a sports center. In 2003, the winning scheme for the Olympic Forest Park was proposed by the Tsinghua Urban Planning and Design Institute, the team led by Chinese landscape architect Jie Hu, with the concept of Olympic Green.

The former land used is a reserve land compared with other contaminated post-industrial sites in North America and Germany, and it is transformed into a country park using the shan-shui structure through the opportunity of a significant urban event. Hence, the 2008 Olympic Games became a catalyst for the redevelopment of suburban areas and the construction of urban green open space, as demonstrated in Figure 69.

Finally, the Olympic Forest Park as one of the new, massive urban projects becomes an instrument to confront today's urban social and ecological problems and to reconcile with China's own past cultural traditions.

The urban society strives to show model solutions for most crowded Chinese cities (Belle 2008). To a certain extent, the large green area relieves the scanty situation of urban green space from the quantitative perspective. In terms of urban social demands, there is a spatial transition from the park's south to north: artificial, semi-natural to natural spaces. Primarily, the southern part is the venue with provisional facilities providing for various events during the games. Afterwards, they continue to be used as recreational and educational facilities for urban residents and visitors. The northern part serves as ecological conservation.

**Figure 69.** The reorganization of urban landscape owing to 2008 Beijing Olympic Games. Source: Photo by author.

With respect to the urban ecology, the shan-shui park applies to modern technologies, such as a hydrological and water quality simulation process and compound water treatment system to enhance the sustainable circulation and utilization of water resources by making use of reclaimed water as the source for the water system and replenishment source for landscape water (Hu 2020). Hence, a self-sustaining and self-regulating water system is formed by planners and designers, particularly in the droughty Beijing City. Moreoveor, a primary ecological corridor is constructed over the fifth ring road according to the ecological principles. The built corridor is the connected part between the northern and southern parts and serves as the pathway for the movement of energy, animals, and people.

The planning and design of the traditional culture incorporate China's history through a grand axis connecting many historic spaces with Olympic Forest Park, a symbolic image of a Chinese dragon turned into the aerial view of a planned stream, and the established shan-shui structure considering the shan-shui culture.

The idea of "Axis to Nature" by the Tsinghua team was presented to guide a spatial transition from the urban historic center to urban nature. It symbolizes the co-existence of nature and humans in an ideal, harmonious manner. The grand spatial axis is planned along the north–south imperial central axis and extended further toward the Olympic Green. On this axis, several essential points are marked, such as the square, the Forbidden City, and tower, forming a spatial sequence in Figure 70. The Tsinghua team reported that the axis has witnessed the changes in the history of Beijing and has carried the symbol and memory of history, culture, and politics. On its ending node, the Olympic Forest Park is precisely arranged. The position of the Olympic Forest Park is identified in "a hierarchical procession" (Selugga 2008). Forbidden City is positioned in a central place as an old cultural symbol.

**Figure 70.** Olympic Forest Park located at the northern end of the Olympic Tower along Beijing's historic urban axis. Source: Photo by ©Ziyue Wang 2021; used with permission.

However, this Beijing historic cultural axis merely shapes a connection between the new Olympic Green and the ancient core area in a relatively simple and direct way. The overall spatial structure of country parks is mostly reflected in the north–south axis and the five-ring-road system. Planners and designers attempted to use the axis to symbolize a formal continuation of Chinese traditional culture. In this situation, the image of the dragon pulse is formed since the dragon is regarded as a symbolic icon for ancient emperors. Therefore, a dragon-shaped water system is planned and designed, as illustrated in the master plan of the Tsinghua team. Thus, an abstract cultural meaning is still essential for the planning and design of Chinese urban landscapes.

Moreover, the axis to nature also influences the shan-shui structure of the park. The man-made mountain range, piled up as the park's highest point, becomes a symbolic terminus of the axis in this spatial framework. The highest artificial mountain and the dragon-shaped water system commonly comprise the shan-shui structure of Olympic Forest Park.

Beijing Olympic Forest Park is considered one of the most important landscape-based projects since it was not only propelled by the international Olympic Games but also sought to plan and design a contemporary urban landscape from both ecological and cultural perspectives. Moreover, Chinese authorities and professionals begin to contribute to implementing urban development programs in the suburban area through the landscape role.

Shougang Industrial Heritage Park

The main site of Shougang Group, established in 1919, is located on the banks of the Yongding River in the west of Beijing. In history, it contributed significantly to the economic development of China. With the advancement of technology after the reform and opening up and the ecological protection promoted as support for the bid to host the Olympic Games in 2008, however, the whole plant ended up with closure and relocation. Then, in 2022, it was repurposed into the office of the Beijing Winter Olympic Games Organizing Committee as well as the venue to hold major events. This presents a new opportunity for functional transformation. According to Chinese landscape architect Yufan Zhu, an "industrial museum" has been built in Beijing on the premise of a previous complex, gigantic system comprised of buildings, equipment, railways, roads, conveyor belts, pipes and chimneys (Zhu and Meng 2016).

Back in 2009, the new round of urban design, with the official name "New Shougang High-end Industrial Comprehensive Service Area", was started. With its independent functional composition and typical features of the traditional steel industry, this closed and managed "compound" ("大院" in Chinese) opened up to the public for connection with the city on all fronts, including transportation, function, visual field, and landscape ecology. Through integration into urban life, it has a promoting effect on the urban environment (ibid.). In this sense, the Chinese landscape project with its own cultural essence became more open, integrated, and inclusive, which conforms to the international landscape trends.

In this circumstance, Shougang Industrial Heritage Park emerged, as shown in Figure 71. In the process of structural transformation, it demonstrates how landscape has been applied to the adjustment of the regional structure, the integration of green space resources, and the restoration of industrial wasteland. According to the master plan drawn up for Beijing, this project is situated at the intersection of Chang'an Street and the east–west axis along the line, within the western green ecological development zone. Close to the inner greenbelt of country parks, it diversifies the form and features of Beijing country parks from the perspectives of shan-shui spaces and industrial remains.

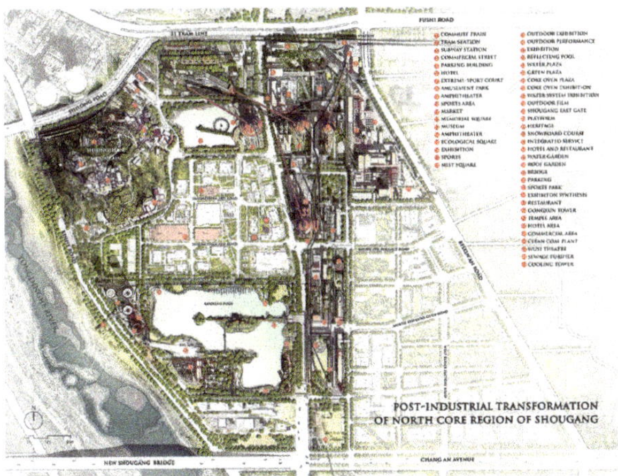



**Figure 71.** Master plan of Shougang Industrial Heritage Park, Tsinghua Urban Planning and Design Institute, 2016. Source: Figure courtesy of ©Tsinghua Urban Planning and Design Institute, Zhu Yufan Studio; used with permission.

It plays a significant role in creating green spaces in the core functional area of the capital, showing the integrated landscape features of both industrial heritage and urban nature. Through a combination with various natural mountain–water resources such as the Yongding River, Shijingshan, and the traditional royal garden system of Three Hills and Five Gardens, the Shougang area is transformed into a landscape ecological park according to the north–south landscape axis, which improves the green space network of parks at a local level.

Located in the north of the Shougang area, the landscape scheme of Shougang Industrial Heritage Park (291 hectares) was put forward by the Tsinghua Urban Planning and Design Institute, which is purposed to find the direction of urban regeneration exploration which revolves around green ecology, functional reuse, and cultural reshaping. Ultimately, the site was transformed through such strategies as the identification, classification, and preservation of industrial remains, palimpsest and weaving of site fragments, etc.

A group of landscape architects with both imagination and creativity made an attempt to demonstrate the potential compatibility of landscape architecture with the post-modernity theories and to achieve their transformation through the design of a post-industrial landscape. This pilot mega-project presents them with an opportunity to explore a new path to building a Chinese large-scale urban park on derelict lands, the significance of landscape architecture in instilling holistic thinking into the intricate and diverse issues of post-industrial transformation from a professional perspective, as well as the possibility that landscape architecture contributes to the intervention in urban regeneration in a wider sense.

Moreover, designers prompted the imagination of this large contaminated industrial area in Beijing, which is based on the understanding of the original site and the plant history over the past century. Landscape architects apply the idea of postmodernism focusing on juxtaposition and horizontality for the shan-shui park in the field of park design. Heterogeneous information is concentrated on the site simultaneously, producing a unique post-industrial image (Lü and Zhu 2020). Subject to influence from the concept of the Collage City in the West, the collage approach relates to the complex state of Shougang Group that has undergone three crucial stages: industrialization, gardening, and regeneration.

Consequently, some industrial buildings and structures remain on the site, showing the ancient gardening style of the Chinese traditional landscape, along with the mega-structures erected for the Beijing Winter Olympic Games, as illustrated in Figure 72. A combination of traditional, modern, and current elements is put on display, presenting both a challenge and opportunity for Chinese landscape architects. The complexity reflects the interrelationship between industrial remains, shan-shui culture, and gardening, as well as urban and national events in time and space. In the urban environment, the complexity may also result from the recognition given to the systematic value of industrial remains, the land and water pollution issue, the lack of vegetation, etc. In this sense, a question is raised by Chinese landscape architects: can new possibilities be explored in the face of such large-scale, hyper-complex industrial remains where neither industrial archaeology nor industrial naturalism is applicable to address their particular complexity of multifaceted superimposition? (Zhu and Meng 2016).

**Figure 72.** The site complexity originating from the overlay of industrial remains, traditional gardening and huge modern elements. Source: Photo by ©Bin Li 2015 (top) and author (bottom); used with permission.

To answer this question, Yufan Zhu proposed the keyword "system". On the premise of protecting the monoliths, connections, and organization of the industrial heritage, the site information is sorted out along the "line" of industrial production processes. Apart from that, in order to activate the spatial potential while creating new economic, social, and ecological values, the original spatial structure is maintained in the process of transformation, demolition, replacement, and integration. A distinctive approach to the adaptive design and atmosphere of the site is taken by designers to achieve large-scale park construction. In the planning of German structuralistic parks, both information and the critical thinking, rational judgement, and rich creativity of designers are required to resolve the complexity of a site and unleash its characters. In this sense, Shougang Industrial Heritage Park is quite inspiring.

From the angle of the shan-shui structure, there is a landscape pattern called "Two Lakes and One Mountain". Figure 73 shows the association between one lake named Xiuchi Pool and Shijingshan Mountain. Moreover, landscape architects interpret the relationship between the other Qunming Lake and Shijingshan Mountain as the traditional garden model of Kunming Lake and Wanshou Mountain in the Summer Palace. From the perspectives of space and scale, the shan-shui structure depicted in Chinese shan-shui culture is inherited on a continued basis. The modernity of the shan-shui city is under consideration. It is expected that the expressions in Chinese large-scale urban parks will be further diversified.

**Figure 73.** The shan-shui structure shaped by the park. Source: Photo by author.

In terms of green infrastructure, the Shougang Industrial Heritage Park was positioned as an urban regeneration and restoration project in line with the guidance of top-down ecological ideas. In the preliminary process of planning, the Beijing Municipal Institute of City Planning and Design and the Shougang Group collaborated in drawing up a comprehensive low-carbon urban development and ecological plan, which led to the first "C40 positive climate development project" in China. Under the concept of ecological sustainable development, it aims to demonstrate the capability of cities to develop in a climate-friendly manner and reduce carbon emissions. A wide variety of elements such as green buildings, clean energy, waste management, water resources, green space, and industrial sites have played a vital role in revitalizing industrial derelict areas, as shown in Figure 74.

**Figure 74.** Low-impact rainwater gardens of Shougang Industrial Heritage Park in the idea of sponge city. Source: Photo by author.

As one of the most successful urban regeneration projects in China, the relative ecological methods of the sponge city can be found for the construction of a resilient landscape on the site scale. To deal with the intricate site information, the multi-layered planning approach of landscape analysis has already been adopted by Chinese landscape architects in a better way. The overlapping landscape system is capable of performing such functions as land rehabilitation, green open space, site drainage, and transportation. The structuralistic syntaxes put forward by Peter Latz in German landscape architecture likely inspired most Chinese landscape architects. For designers, the paradigm of structural analysis promotes the organization of such large-scale sites. However, Shougang Industrial Heritage Park is definitely distinct from Peter Latz's Landscape Park of Duisburg Nord. In China, large-scale urban parks are constructed in their own way. This project clearly reflects the attitude of Chinese landscape architects regarding the abandoned industrial site in the course of urban regeneration, the urban–industrial nature of artificiality and wildness, and the modernity of the shan-shui structure.

Currently, there are various attributes shown by Shougang Industrial Heritage Park, such as the urbanity reflected in the daily lives of residents and tournaments and conferences for city, the cultural nature of shan-shui, the complexity of perception, and the conflicting nature of past and present. The integration of attributes contributes to the inclusiveness of this distinctive shan-shui park. In the future, the development of a compound post-industrial landscape system will continue with the evolution of space, place, and nature.

# 5. Systematic Comparison and Reflection

"Think global, act local." (Geddes 1915)

It is within the regional cultural concepts of urban landscapes that these two advanced large-scale urban park models in the West are compared. On this basis, this chapter presents the revealed similarities and differences for further reflection on Chinese shan-shui parks. Their parallels are deduced to solve the research question about how to view contemporary large-scale parks for adaptation to the ever-changing conditions in terms of urban spatial structure, society, and ecology.

By adopting critical rationalism approaches, two distinctive park models give priority to the understanding of landscape over park at the urban level. They are considered not only as a strategy for site renewal and transformation in post-industrial society but also as eco-machines for natural processes, to deal with the relationships with revised cities and urban nature. A discussion is conducted about their differences at first from the perspective of urban landscapes within two different theoretical schools of *landscape urbanism* and *landscape structuralism*, which is followed by what is stated explicitly at the level of large-scale parks through the questions raised by the author.

## 5.1. Parallels

In general, both large-scale park models reflect the advancement of contemporary landscape architecture that has experienced vigorous growth over the past fifteen to twenty years. Undeniably, its significant growth in both research and practices is coupled not only with the changes occurring to the urban spatial structure, social transition, and ecological challenge but also the growing demands for the qualitative development of urban green open space. Accordingly, two kinds of large-scale urban parks emerge for analysis as the specific forms of urban landscape with unique cultural identities. Through a review of their explanations in previous chapters, their parallels are summarized into seven points.

1. Critical Rationalism Approach

   The critical research approach is the first notable aspect of their similarities. As explained in Chapter 2, there are two specific critical approaches taken to explore contemporary urban landscapes. One is the North American *critical thinking* proposed by James Corner and the other is the German *critical structuralism* interpreted by Peter Latz. In line with Karl Popper's principle of "falsification" in scientific theories, they can be summarized into critical rationalism approaches. The approach contributes to a critical exploration of two large-scale park models compared with classic pastoral nineteenth-century parks. They demonstrate a commonly critical attitude towards contemporary urban landscapes. Additionally, it offers guidance for the further rethinking of the Chinese urban landscape and its country parks from a methodological perspective.

2.    New Landscapes at the Urban Level

The second similarity relates to the comprehension of two large-scale urban parks. Firstly, an expanded scale is involved in the parks. The dissolved urban spatial structure and the expansion of urbanity into landscape lead to the extension of parks into the whole urban regions. Secondly, two large-scale parks are viewed as open, extensive, and interconnected landscape at the urban level, that is, urban landscapes, despite "park" as a keyword. It transcends the concept of the park. From this transition, it can be seen that the landscape has become the focal point of consideration and planning of urban development. Meanwhile, the transition played a role in reversing the ideal interpretation of the park from a pastoral perspective. Regardless of how large-scale parks are envisioned, they are not restricted to a harmonious image of painted landscapes. Thus, a non-pastoral perspective could be shaped by more creative conceptions. It reflects the evolving park concept in respect of landscape architecture.

3.    Design Paradigm with Cultural Identities

The third similarity implies the intellectual search for the identity of parks. In comparison with traditional parks, two large-scale urban parks represent the intellectual constructs in specific cultural contexts, which leads to the different ways of park models development. Allowing for the transition of society, ecology, and space in urban reality, it is unrealistic to position present-day large-scale parks in a wholly ideal and romantic dimension. For them, rational and critical analyses toward different dimensions are required, such as the North American large parks with the organic identity, or the German large parks with the structural identity. As for the issue of how their cultural identity is shaped, a discussion will be conducted in the section on differences. As a crucial element of park models, the identity prompts Chinese shan-shui parks into exploring and determining their own characteristics in the future based on the *cultural imagination* and creativity of designers. This step may be consistent with the shan-shui park design approach to be reflected. The coherence shows similarity to that between approach and identity in North American and German models. The explicit park identity contributes to its conceptual approach.

4.    As an Instrument for Site Transformation

The structural change in post-industrial society has led to the emergence of countless abandoned sites in cites. Meanwhile, the transformation of derelict industrial sites has been accepted as the theme of large-scale park development in most cases, accounting for the recognition given to the significant role that landscape plays in urban regeneration and renewal. Routinely, large-scale urban parks are viewed as a means to achieve site renewal and redevelopment sustainably. It is demonstrated that clues can be provided by the research on two large-scale park models for the Chinese shan-shui parks in the transition of Chinese society. This will be discussed in the part of changing city as an urban phenomenon.

5.  As Eco-machines for Processes

Eco-machines, a term referred to by others as landscape machines or living machines, are applied to describe both large-scale urban park models from the perspective of processes. Not mentioned in previous chapters, the concept will be explained in this section. The views of this study are close to the interpretation made by Dutch landscape architect Paul A. Roncken's interpretation:

> Eco-machines are "made of landscape features and are driven by landscape processes, and in the meantime, they produce a multitude of food products, natural biotopes, clean air, clean soils and so on." The priority of eco-machines is "not only to protect and understand nature but also to feed those processes that sustain nature's resilience and thereby harvest all the by-products and spin-off effects that we need as human beings." (Roncken et al. 2011)

Eco-machines are aimed at presenting a relationship between two large-scale urban parks and ecology, rather than calculating the specific inputs and outputs. Like machines, both expansive parks fulfill their functions in the ecological processes created and maintained through technologies when there is a harmony between technologies and nature. According to ecological rules, the natural processes in large parks are intended not only for preserving natural elements and resources, restoring and improving ecological environment, but also for enhancing ecological resilience in the face of disturbance. Therefore, they are eco-machines, not "environmental cleaning machines" (Meyer 2008).

6.  Dynamic Parks with Revised Cities

The association between park and city becomes increasingly close. Two large-scale parks are constructed in *urban regions*, usually with the changing nature of the revised city, as reflected in the dissolution of dominating urban organizational form and the transformation into a post-industrial society. There is an essential foundation laid for large-scale park concept development by the ever-changing urban environment and the critical reconstruction of urban landscapes.

Moreover, their close link demonstrates "how the city is to be viewed" (Cranz 1982). The large-scale park concepts are determined by certain visions of the city. In the third chapter, the renewed urbanism, especially the *landscape urbanism* program in North America, was discussed. It is a form of conceptual city model where the city is considered as what is similar to an organism containing fluid, substance, and energy, and has relationships. Through an organic approach, the organic insight into the city is linked to the concept of large parks. Likewise, Thomas Sieverts conceptualized a new urban model of in-between-city in Germany. In Chapter 2, a "city-landscape continuum" was explained, suggesting that the city is neither an entity nor an organism from the perspective of its immanent "difference" and "diversity". This is premised on the heterogeneity of city life, giving rise to the German concept of large park space with qualitative characteristics. Both visions of the city establish relationships with large-scale urban parks.

7.    Dynamic Parks with Urban Nature

Due to the shift in the relationship between nature and the city, nature is closely related to expansive parks, as discovered in North American and German academic circles. According to Charles Waldheim, ecological forces and flows represent part of the city. The integration of nature and the city was supported in his renewed urbanism model, as described in Chapter 3. Similarly, Peter Latz argued that to safeguard urban landscapes as basic life resources now and in the future, it is necessary to address the irreconcilable contrast between the city, nature, and technology (Weilacher 2008). Regardless of different cultural interpretations and images of nature in two developed regions, nature ceases to be the scenery in contemporary urban conditions. Reclaimed and regenerated by means of technologies, urban nature is determined by urban residents. Urban nature is inextricably associated with two large-scale urban parks needing to be considered by planners and designers.

In a broad sense, the understanding of urban wildness is of practical significance, as reflected in the planning and design of the two large-scale parks. Especially, self-organization of resilient nature is cultivated by urban wildness through technologies. In the meantime, the interaction between nature and the urban environment is enhanced in a distinctive organized way when dynamic processes of ecosystems are revealed. Therefore, urban wildness is integrated with purpose into two large-scale parks on post-industrial sites. For Peter Latz, urban wildness enables the coincidence of nature. For James Corner, it symbolizes the expansion of the living ecosystem into the urban field. Both of them demonstrate the demand for urban wildness.

*5.2. Differences*

Relative to the two large-scale urban parks on derelict lands in Germany and North America, their striking differences are elaborated on in this section. The ideas of German large parks originated from Peter Latz within the scope of the *landscape structuralism* school of thought, while those of North America originated from James Corner within the scope of the *landscape urbanism* school of thought.

5.2.1. Urban Landscape Formulations

In Chapter 3, an analysis was conducted regarding the reconstructed ideas of contemporary urban landscapes in North America and Germany. Prior to a clear comparison of the two expansive parks, there are three vital aspects primarily discerned, which are landscape understanding, landscape and ecology, and landscape and urban life. It shows the research logic of comparing urban landscapes at first and then large-scale parks.

Landscape Understanding: Coherent and Creative

In developed countries, the landscape has been accepted as the central point through the theoretical and practical analyses of two large-scale parks. There is a possibility that this cognition of German landscape architecture is reflected in the projecting of structuralistic parks, primarily with Duisburg-Nord Landscape Park as the starting point. Over time, a coherent and contiguous understanding of landscape is gained along the thread of the German cultural landscape.

The reasonable reference of the long-standing site *information* may be the origin of coherent German landscape understanding, that is, the objective existence in urban regions. On the specific site, the relatively stable and flexible structure is likely to be the premise of a selective introduction of complex and describable *information* into German structuralistic parks. In other words, it is possible to subjectively analyze, conceive and establish the spatial structure by having reference to the physical geographical site, such as regional texture with inseparable landscape elements with unique sociocultural characteristics.

The site-specific *information* offers strong support for further conception and expression. As mentioned above, the process of reference is the ongoing "decision-making process" for planners and designers (Latz 2008a). For professionals, it is necessary to find and discern "what force the existing objects already have, what density of information they already possess and what density of information first has to be introduced into the project", as quoted in Udo Weilacher's (1996) book *Between Landscape Architecture and Land Art*. A conclusion drawn in the following part is that German large-scale parks rely on *information*.

To sum up, the German landscape from the 'structuralistic' perspective is viewed as a spatial structure that represents the core of the landscape. Regarding the idea of heavy reliance on site-specific morphology and elements, it may be closely associated with the recognition given to the German cultural landscape as the physical object influenced by the geographic idea of landscape.

Differently, the understanding of North American landscape architecture is influenced by the great ambitions of landscape shaping urban space and a strong sense of change when potentially creative thinking functions. As a result, there are emerging ideas applied to critically review and replace certain traditional ideas. The situation shows similarity to the evolutionary process of proceeding in negation. In this cultural context, the North American landscape is viewed as a "verb", as "process or activity" (Corner 1999). It lays emphasis on "the effects of constructed landscape in time" and "how it works and what it does" (ibid.).

The comparison contributes to the understanding of landscape between German and North America in a coherent way versus a creative way. There are two distinct ways of treating landscape in keeping with the elements on which large-scale urban parks on post-industrial sites intend to rely. They also constitute the second question of comparing two expansive parks.

- German landscape understanding in a coherent way, associated with large parks relying on *information*.
- North American landscape understanding in a creative way, associated with large parks relying on *imagination*.

Landscape and Ecology: Representation and Metaphor

Currently, it is essential to integrate the construed nature in cultures with contemporary cities and landscapes. It is possible for the cultural image of nature to exist in the field of landscape architecture. In different cultures and at different times, there are multiple ecological ideas arising from the practice of learning nature, which is through the ongoing interpretations of nature by professionals. In this combination, the attitude toward German ecology undoubtedly tends to be the artistic interpretation and representation of nature. However, ecology is first regarded as a metaphor for the city dominated by landscape in the current North American academe. Along with landscape, ecology is treated as an "agent of creativity" based on "highly interactive processes and relationships" (Corner 1997).

With respect to the German landscape, nature and culture are considered a "continuum" (Latz 2008a), not possessing an opposite relationship. The understanding of nature shall be gained in a unique cultural condition that may change with society:

> "Landscape architectural design always—whether consciously or not—deals with society's position towards nature. Thus, it makes reference to an inherent paradigm that has guided landscape architecture for centuries. Today, it no longer suffices to consider nature in isolation, as an antipode to cultural creation. If survival on earth is to be safeguarded for the future, technical and natural phenomena, culture and nature must be comprehended as a unit, a continuum." (ibid.)

According to Peter Latz, those newly emerging open spaces within the urban contexts in German post-industrial society, such as city parks and gardens, were expected to symbolize nature and landscape. Artifacts (technological structures or elements) are used as symbols of nature and life in nature (ibid.).

In the analyses conducted on the contemporary North American urban landscape in the 1990s, the metaphor, especially in an ecological sense, shows its essential orientation. This metaphor contributes to a specific set of ecological ideas as distinct spatial generators, driving the redevelopment of large-scale landscape architecture. It makes "ecology not a remote 'nature', but more integrative 'soft system'—fluid, pliant, adaptive fields that are responsive and evolving" (Corner and Hirsch 2014). In the planning and design of large-scale urban projects, the understanding of urban living surfaces has played a crucial role. Ecology itself becomes "an extremely useful lens through which to analyze and project alternative urban futures" (Corner 2006).

Landscape and Life: Diversity and Unpredictability

German landscape attaches much importance to the diversity of social life as a foundation for the generation of diverse urban space. Landscape is viewed as an urban green open space, guaranteeing a free and diverse urban life.

By contrast, there may be more unpredictable factors or unpredictability in urban life incorporated into its concept of landscape in the existing North American landscape. This shift indicates that a deterministic inference in the large park concept exerts almost no influence. For James Corner, life is the origin of unpredictability.

He considered life "as both a specific and autonomous system of networks, forces, combinations, unfoldings, events, and transformations" (Corner 1997). Since the understanding of life requires creativity for the urban landscape, landscape as a condition is set up for an uncertain life to unfold and evolve.

In summary, there are three levels of comparison in the urban landscape with their key bullets, as shown in Table 4 below.

**Table 4.** Comparison of contemporary urban landscapes in two theoretical conceptions at three levels.

| Difference | Germany | North America |
|---|---|---|
| At the Level of Contemporary Urban Landscapes | Landscape structuralism | Landscape urbanism |
| Landscape Understanding | 1. In a coherent way<br>2. Landscape as a spatial structure, the relatively stable structure becoming the immanent core of landscape | 1. In a creative way<br>2. Landscape as "process or activity" |
| Landscape and Ecology | Ecology as the artistic interpretation and representation of nature | Ecology as a metaphor, and ecology and landscape as "agents of creativity" |
| Landscape and Life | Life's diversity: landscape as urban green open space guaranteeing free and diverse urban life | Life's unpredictability: landscape as the conditions set up for uncertain life to unfold and evolve |

Source: Author's compilation based on data from Latz 2008b; Corner 1997; Corner 1999.

## 5.2.2. Large-Scale Urban Parks Conceptions

In this section, there are eight different aspects showing the disparities between expansive parks in North America and Germany, which provide some conflicting answers to several in-depth questions about the two models as follows. The relative contents are summarized in Table A5.

- What is the cultural identity of each park paradigm?
  The structural identity vs. the organic identity
- What elements are relied on for their conceptions at the beginning of thinking?
  Relying on information vs. relying on imagination
- What kinds of techniques are employed to conduct their conceptions?
  Objective representation technique vs. imaging techniques
- What are their most critical contents in large-scale park planning and design?
  Shaping structural space vs. establishing fluid, adaptive field
- What are the aims of developing spaces in two large-scale parks?
  Spatial qualities vs. spatial performance
- How are natural processes in two large-scale parks regarded?
  Cultivated process of nature vs. productive process of nature

- What kinds of elements are conceived in large-scale park planning and design?
  Site-specific elements vs. non-site elements
- How is a series of qualitative characteristics of large-scale park models interpreted?
  Qualitative characteristics of the German model vs. those of the North American model

## Structuralistic and Organic

This research elaborates on two distinctive models constructed as park paradigm shifts. The German large park is deemed the structuralistic park paradigm, whereas the North American one is regarded as the organic park paradigm. The two models are contributory to the contemporary development of park cultural identities. To explain similarity, it is necessary to set up the cultural identity of two large-scale parks through intellectual thinking.

Based on ecosystem dynamics, the organic identity is derived from the emerging ecological ideas and principles of ecosystems transformed into North American large park concepts. Meanwhile, distance is maintained from the sophisticated science of a balance of nature in a critical way. With the organic approach adopted, it is subject to the influence of the disciplinary framework of landscape urbanism. Especially, it reflects the combination of landscape, ecology, and engineering. As indicated by the understanding of organic, North American landscape architects believe that North American large parks are the organic infrastructure essential for enhancing the ecological functionality of the living urban surface. Driven by the functional understanding of nature, North American large parks evolve toward the organic model.

As for the structural identity, it is premised on Peter Latz's unique design philosophy of "decoding", understanding and re-interpretation, as well as "new syntax of landscape" (Weilacher 2008). "The landscape of the park is a rational construct whose layers of information contained in its structure were preserved and transformed" (Latz 2013a). The "structure" is deemed indispensable because of its significance valued by Peter Latz. He even described it as "robust" and "fascinating" (ibid.). As suggested by the understanding of structural, the derelict industrial land can be reorganized and transformed reasonably under the inherent material and spatial connections, without affecting features on the specific site. It has influenced a generation of German landscape architects profoundly.

## Information and Imagination

With regard to the planning and design of two large-scale parks, there are starkly different ways on which Peter Latz and James Corner rely. Peter Latz relies on information to analyze or interpret the physical site and its context. The concept of information is aligned with the coherent understanding of landscape. Differently, James Corner presents an "eidetic scope of landscape creativity" to primarily conceive of sites according to his idea of "imagination" (Corner 1999). The two aspects are regarded as different starting points of thinking.

In Peter Latz's thinking of the structuralistic approach, the reliance on accumulated *information* in the history is a foundation for his idea of cultural contextualization. Then, the designed site is integrated with the interpretation of designers to shape spatial forms and establish spatial connections in the course of exploring the palimpsest of the site. As argued by Peter Latz, the design process is an "invention" of informational layers overlapping with existing systems, before designers give consideration to shape or expression at all (Latz 2008a). Without the fundamental information in the external reality, it is difficult to envision Peter Latz's analysis and further conception of German large parks.

However, there is another starting point of thinking in North American large park planning and design, that is, subjective imagination. It is consistent with the creative understanding of landscape. Due to the profound influence of J. B. Jackson's innovative understanding of the North American landscape and postmodern cognition of space, the *cultural imagination* of James Corner is established as the cultural embedding, as described in the North American large park chapter. In his view, imagination is "a power of consciousness that transcends visualization" (Corner 1999). It shows an "eidetic and subjective way" (ibid.).

Representation and Imaging

For large-scale parks, the conceptions are materialized through the application of their own representational techniques. Naturally, these techniques are Peter Latz's objective representation and James Corner's "imaging techniques". The two modes of representation are pertinent to the *information* as described in opposition to imagination.

With reliance on *information*, Peter Latz adopts objective representation to shape the structural space with purpose. As implied by this technique, objective existence is the foundation of the subsequent German structuralistic park conception. In essence, the objective representation represents an interaction between objects and subjects, or between landscape and planners or designers to be exact. The process of representation is viewed as objectifying subjects or objectification. The technique does not necessarily indicate the emergence of uniformly conceptualized results, which is because the selection and handling of a variety of visible and invisible *information* are clearly different from the perspective of Peter Latz with more or less individual creativity. "We select some information from the surroundings and make an idea in our head. Each person has another method to combine the information. There are different information layers, and you may understand only one or two, but somebody else may understand 50", as quoted by Arthur Lubow in "The Anti-Olmsted" (Lubow 2004). There might be no problem arising from the objective representation. On the contrary, what matters most is the active understanding of objects in one's mind.

However, the objective representation has something wrong, as affirmed by James Corner. His consideration of representation appears to be radical, particularly for the discovery of the faint effect of creativity in landscape architecture. For this reason, he directs criticism at the objective representation without fully producing an active role, rejecting the continued application of it in the conception of the North American large park. Through imagination, he stressed that representation should be closely related to "a mental conception" (Corner 1999) and focused more on improving and creating various techniques of representation. In his view, it is beneficial to apply the imaging techniques of conceptualization, such as "mapping, planning, diagramming, and sectioning" that free the designer and planner from the restriction of representation (ibid.).

Spaces and Fields

The central point of two large-scale park conceptions is to construct either space or field. Among the most crucial contents in the German large park is the creation of structural space, while the establishment of a fluid and adaptive field plays an equally important role in the North American organic park.

In German landscape architecture, the reflection and emphasis on the concept of space are elaborated on by Henri Lefèbvre in the 1974 inference of "social production of space" with the spatial turn. Considered a social construct, the space is "produced and reproduced through human activity" (Lefèbvre [1974] 1991). German structuralistc parks as urban spaces are prepared for social appropriation in everyday life. These insights lead to the formation and development of specific and differentiated spaces and spatial forms. The shaping of space is integrated with Peter Latz's landscape much more "as a spatial structure of informational layers shaped by people that develops permanently and dynamically" (Weilacher 2008). Therefore, it is the core of German structuralistic parks for the shaping of structural space.

North American organic parks are not intended to shape space or spatial material form. Instead, they are dedicated to setting up the "fluid, pliant fields" able to "absorb, transform, and exchange information with their surroundings" (Corner and Hirsch 2014). The "fields" are equivalent to the complex systems as discussed in the third chapter. They are extensive and interconnected networks where more attention is drawn to the interrelationships between things in space and the effects produced through such dynamic interactions than to the solely compositional arrangements of objects and surfaces. Ultimately, the fluid field of North American organic parks is constructed not only for adaptation to unpredictable changes and desires but also for "new forms and combinations of life to emerge" (Corner and Allen 2001).

Qualities and Performance

On the basis of shaping space in opposition to field construction, German structuralistic parks develop their spatial qualities further through the structure. Differently, North American organic parks rely on the functioning framework or matrix to produce their spatial efficacy or effectiveness, that is, performance.

Peter Latz showed aspiration to qualitatively develop space in all its facets and dimensions. He emphasizes the possibility of a close association between the spatial qualities of German large parks and the analysis of Henri Lefèbvre on the urban landscape and the "difference". Henri Lefèbvre provides a significant insight that the European urban landscape should be defined at the "specifically urban level", in comparison with the levels of superstructure and infrastructure (Lefèbvre [1970] 2003). At the three levels, a specific urban space is maintained by German structuralistic parks, with the coincidence of diverse and free urban life and natural process. As the organic infrastructure, North American large parks are understood through functions and efficacy. Therefore, the spatial characteristic is enhanced by the spatial qualities as developed by the concrete spatial forms of urban life and urban nature. In contrast, there is no association shown by the formation of spatial features with the infrastructural functions more connected to ecological measurements and technologies, despite a role played by them through ecologically spatial patterns, such as pathways, corridors, edges, patches, and matrices, as well as their relationships.

The organic functions performed by North American organic parks focus on the "formative effects of landscape in time" (Corner 1999), that is, the landscape *performance* linked to the process. Referring to the ecologically spatial patterns, the "formative" is compared with the specific characteristic forms in the German structuralistic park. For North America, its significance is less reflected in these known conceptual patterns than in the course of producing the effectiveness via dynamic interactions of patterns over time. The effectiveness reflects the key issue encountered by North American landscape architects, that is, how large parks work. To support the spatial performance alternative to qualities, James Corner opposed the exclusive emphasis laid on the formal and visual qualities of landscape since the landscape can be changed into a "dead event" by the priority for those alone (ibid.). In this sense, its spatial qualities and forms are not the priority for the North American organic park spatial performance.

Cultivated and Productive

Both large-scale urban park models, as admitting dynamic processes in the non-equilibrium paradigm of ecosystems, require the understanding of the nature-related process. However, there is a clear difference between them.

Regarding German structuralistc parks, the cultivated process of nature is purposed to strike a fine balance between the "untouched" and the "built" (Latz 2003). While accepting a fragmented world, Peter Latz allows room for the coincidence of nature in the web of the layout (ibid.). Biotopes of nature experience autonomous growth, maturity, and even decay in German large parks. However, the cultivated process is not limited to the natural process initiated and sustained by technologies according to the rules of ecology. Pertinent to the artistic representation of nature by artifacts as a symbol, it also embodies the understanding of landscape and ecology among Germans as discussed above.

The cultivated process marks an attempt made to keep and recycle, rather than generating or creating something. On the contrary, North American large parks are purposed to achieve the efficacy of nature, which is based on the functional understanding of nature and the focus on park performance. As suggested by the productive process, the constructed urban nature provides a driving force for the development of North American organic parks in circumstances, making them effective through seeding, staging, and ecological succession (e.g., James Corner's Freshkills Park project) in a large part. It ends up with an evolving, open, resilient system for the large park as a response to the change of needs and desires.

## Site-Specific and Non-Site

In terms of German structuralistic parks, site-specific meaningful elements are made use of to form an open and multilayered landscape structure in the practice of planning and design. Extracted from the sites in concrete urban conditions, they are regarded as structural elements by Peter Latz. According to Udo Weilacher, these elements relate to topographic and hydrological morphology, transportation systems, building structures, open space systems, and additional relevant networks. The site-specific elements are crucial to the spatial structure and form, on account of the coherent understanding of landscape and reliance on *information* for analysis and design.

With regard to North American organic parks, an open-ended multilayered framework or *matrix* consists of non-site elements as conceptual diagrams. Mostly associated with ecologically spatial patterns, these diagrams are transformed into conceptual elements in planning and design. As for the understanding of non-site elements, it may be influenced by Robert Smithson, who refers to the non-site as "an abstraction of a physical geographical site that can represent the site without the need to resemble it" (Smithson 1996; Wall and Dring 2015). His idea is used as a reasonable argument for North American large park non-site elements of conceptual diagrams from the landscape-ecological perspective.

## Qualitative Characteristics

Qualitatively, two large-scale park models have been analyzed by using five characteristics: complexity, diversity, sustainability, appropriation, and identity. They are summarized in Table 5 below.

**Table 5.** Five qualitative characteristics for the qualification of two large scale urban parks.

| Qualification | Structuralistic Parks | Organic Parks |
|---|---|---|
| Complexity | Complex design process, the intersection of existence and imagination | Adaptive social and ecological processes with the unpredictability |
| Diversity | Spatial diversity and difference | Heterogeneity in the landscape-ecological meaning |
| Sustainability | The coherent development of space in terms of history, memory, social appropriation and represented nature; the recycling utilization of post-industrial remnants | Resilience in the landscape-ecological meaning |
| Appropriation | Urban spaces prepared for diverse social appropriation | Social self-organization emphasizing programmatic indeterminacy |
| Identity | Structural | Organic |

Source: Table by author.

- Complexity

The complexity of the German model is referred to as the design process where there is an intersection of existence in the material world and imagination in the designer's mind. It is difficult to describe the complex design process one by one as it usually requires the selection, extraction, and re-interpretation of a range of *information* through individual thinking. It is thus difficult to explain which *information* or elements could exert influence on the insight of designers for further conception. It might be based on a designer's understanding of the site that the intricate connections between *information* and the new structure of the park are represented. Under the influence of system theories and ecology, the complexity of the North American model is analyzed directly. The complexity of its organic park is demonstrated through its adaptive social and ecological processes with its unpredictability.

- Diversity

The diversity in the German model arises from the diverse spatial forms of society and life. Under the North American model, it is defined as heterogeneity from the landscape-ecological perspective, according to the research on structure, function, and change in a heterogeneous land area "composed of a cluster of interacting ecosystems" proposed by Richard T. T. Forman and Michel Godron (Forman 1987).

- Sustainability

Similarly, the North American sustainability is explained from the landscape-ecological perspective. It is labeled as the resilience developed as among the essential properties in the 1992 dynamic model proposed by C. S. Holling. In contrast, the German one shows a kind of continuity. This reveals the development of space coherent with the three layers proposed by Peter Latz: interpreted history and memory, social appropriation, and represented nature. Moreover, on the basis of keeping almost everything on the local site, German sustainability is also linked to the recycling utilization of post-industrial remnants.

- Appropriation

The appropriation in the German model continues to follow Peter Latz's urban spaces created for diverse social appropriation suited to different levels of willingness and demands of individuals, groups, communities, and society for different uses in the forms of activities, programs, and events. It reflects the above diversity from a social perspective. However, the appropriation in the North American model requires social self-organization, with programmatic indeterminacy as the emphasis. In the process of diverse, creative, and uncertain urban daily life, the appropriation of space is adjusted to suit specific situations and demands. The significant difference between them clearly reflects the opposite part of landscape and life: diversity versus unpredictability.

- Identity

  The last but still important level is the German structural identity, which is distinct from the North American organic one. It has been discussed in the first question. Each characteristic can be interpreted in its own way, as explained in Chapter 4.

*5.3. Reflections*

Herein, the premise for rethinking and readjustment of the Chinese urban landscape is to conduct a search for its own cultural identity, not simplistic, blind imitation. Given the truly ecological landscape architecture in North America and German-shaped cultural landscape architecture, it is possible to explore the identity of the Chinese urban landscape from the following three perspectives: landscape understanding, landscape and artificial nature, and landscape and diverse life.

### 5.3.1. Expanded Landscape Concept

The landscape concept of Chinese landscape architecture as a profession can be expanded through imagination and creativity on site, with consideration given to the essence of Chinese culture.

Actually, the current cognition of the contemporary Chinese urban landscape can be gained by integrating the modern idea of city beautiful with the ideal and traditional understanding of nature. It arises from the imitation of first nature in a long-standing and pain worldview, the harmonious idea of nature expresses the worship and passion of nature, as well as the reshaping and representation of nature in a man-made and artistic way.

The dominant theoretical comprehension of the Chinese landscape is contributed to by the combination of city beautiful and a traditional view of nature, which leads to homogeneous urban landscapes for the pursuit of excessively superficial decoration and artificiality spreading over Chinese cities. This is coupled with a simplistic and ideal understanding of landscape. They reflect a blind following of the so-called trend in developed regions without critical thinking and an absolute landscape concept without reliance on the changing urban condition.

Therefore, it is necessary to consider expanding the Chinese landscape understanding, which mostly tends to be an ideal image or a symbol with abstract Chinese cultural meanings. Actually, there are more and more Chinese landscape architects viewing landscape as a diverse, specific space, and green infrastructure for the improvement of social and ecological qualities for urban green open spaces.

Regarding planning and design, the landscape is related to the extensive (re)construction and (re)development of urban space. As described in Chapter 2, arguments have been made by the theories of landscape urbanism in North America and "a city-landscape continuum" in Germany. Both demonstrate the increasing roles and potential of contemporary urban landscapes at the urban level. The urban stems from the explained and reiterated conceptualisation of the urban region in the research. In addition to the Western theoretical concepts, the urban-scale landscape planning and design projects are quick to spread nationwide, as indicated by Chinese landscape architect Jie Hu. He proposed to "break out of microscopic scale landscape planning and design; strengthen the category of 'ecology and culture as the guide of landscape planning and design in urban scale" (Hu 2020).

As a result, the Chinese landscape is not regarded as a "city landscape" but an "urban landscape". Since the 1980s, the concept of the "city landscape" has emerged in the Chinese landscape architectural discipline to describe the landscape, but only in and around the city center. However, its understanding ceases to be appropriate for the growing urban spatial structure, whether at present or in the future, which requires its expansion to the urban region. Especially, it is possible for the cognition of the urban landscape or landscape at an urban scale to promote the development of an urban–rural integrated spatial system under the context of the urban–rural divide.

Moreover, the profound influence of traditional gardening techniques has restricted the Chinese landscape to a relatively small scale. The traditional techniques of gardening, such as imaging the big from small and world-in-a-pot, are applied to the close, independent, and confined spaces in both historic northern-royal and southern-private gardens. This deep-rooted historic and cultural factor also constrains the understanding of contemporary Chinese landscape at an urban scale, which widens the gap between contemporary landscape architecture and regional planning. The former and the latter are strictly distinguished as a small scale and an urban, regional scale, respectively. In this circumstance, it is difficult to find or better understand the significance of landscape in transforming and redeveloping wide urban spaces.

Therefore, the integration of Chinese landscape architecture with regional planning is supposed to be supported by adopting a holistic approach suitable for landscape at an urban scale, rather than traditional gardening techniques. As demonstrated by the shan-shui city, the shan-shui structure created by ancient urban planners should be inherited by contemporary urban planners and landscape architects.

In order to avoid monocultures and uniformity, there is a necessity to seek a distinctive topography and biological structure. The development of a diverse structure with ecotones, edges, and appropriate layers is required to encourage a wide variety of plants and animals to various habitats and microclimates. Meanwhile, it is necessary to conserve a post-industrial landscape structure in the planning stage of Chinese inventory urban regeneration and superimpose it on the unique urban fabric. In this way, large-scale urban parks can be regarded as an essential form and planning strategy for the activation and redevelopment of the whole urban region. At the urban level, the power and potential of the landscape are always worth paying attention to for each Chinese landscape architect, because it plays a vital role in the interpretation, superimposition, and multilayering of urban spaces as well as the integration and communication of spatial elements. Moreover, it is necessary to understand the openness of the landscape and natural processes in terms of coincidence, rhythm, and time. These essential contents represent the essence of urban landscapes and parks keeping dynamics, high quality, and adaptability for the construction of resilient cities in the future, whether in the West or in China.

## 5.3.2. Landscape and Artificial Nature

On post-industrial sites, the development of dynamic landscapes has embraced urban nature with the artificial and wild characteristics, which requires Chinese landscape architects to follow the trend of working with nature to foster resilience. The ultimate planning and design are determined by the physical and biological factors of the site and man and his technical muscle. An attempt should be made to restore the natural processes and cycles existing on the park site, rather than disrupting them. As for intervention, it must be restricted to addressing destructive conflict. In this sense, it should remain partly an area of wild, undisturbed urban nature in the course of preserving abandoned industrial sites.

Furthermore, when it comes to the relevant ecological theory in relation to Chinese landscape architecture, it is necessary to fully understand the concept of ecological infrastructure at three levels in theory and practice: region, city, and district. For our current Park City construction, priority shall be given to protecting and strengthening the ecological structure. In fact, the opposition to a rampant, destructive urban exploration, the sustainable knowledge, and an ecological approach provided a unique perspective on making progress in construction and development for China.

Additionally, there are hardly any contemporary ecological ideas combined with Chinese landscape architecture to practice the traditional design philosophy of "unity of man and nature". This is possibly due to the belief in the ideal view of nature that could address almost all contradictions and complex environmental problems. It is difficult for the primitive and ideal ecological idea to function in the contemporary urban environment. In this respect, some inspiration can be derived from the ecologically grounded knowledge in the West to tackle the environmental problem.

Thus, it is crucial to gain a contemporary understanding of nature in the Chinese urban social context. More specifically, it is necessary to form a cultural construe of contemporary nature, exactly according to the North American dynamic, non-equilibrium ecological view or German technical–natural balanced ecological view.

## 5.3.3. Landscape and Diverse Life

The rich connotations of the contemporary landscape can be explored in urban life. As discussed above, the unpredictability of life provides the North American urban landscape with more creative possibilities and potential over time, and the acceptance of diverse urban life contributes to the formation of diverse social spaces in German landscape architecture.

As for the Chinese urban landscape, it is not simplistically explained as an ideal image or a symbol with abstract cultural meanings. To some extent, it is difficult to accept both explanations pertain to the dimension of cultural superstructure, because of the disparity from contemporary urban everyday life. For Chinese landscape architecture, there are more opportunities to make interpretations from contemporary urban social life, if life is regarded as the essential source of planning and design. Meanwhile, to landscape architects themselves, the real diverse life is also where high-quality landscape planning and design originated, which is sometimes ignored by us. As a source of inspiration, the richness of living space perception and experience is significant to landscape architects in terms of spatial design.

The landscape projects should embrace the diversity of life, which can be examined by ordinary people in their daily lives. The evolution of the urban landscape and large-scale urban parks on post-industrial sites is based on the correlations between users, spaces, and activities and events. In each socio-cultural context, there are various and complex spatial atmospheres, experiences, perceptions, and feedback over time.

In summary, uniformity and diversity can be demonstrated at the international and local levels through the commonalities and differences between contemporary urban landscapes and large-scale urban parks. There are a series of vivid images about contemporary dynamic landscapes being superimposed, intertwined and changed in time and space, which is accompanied by various information from urban landscapes, post-industrial landscapes, and large-scale parks that are dismantled, stitched together, and compounded by professionals worldwide. To a certain extent, the cross-cultural research is purposed to present, understand, and reveal the differences and similarities. However, in the present and future, globalization, extensive communication, and collaboration may blur these differences and make them less distinguishable. This is definitely a great challenge for landscape architects. Alternatively, more points of difference will be referred to for different landscape understanding. Instead of being limited to an analysis of the vast differences reflected in these urban landscapes, this study provides in-depth insights into the diverse and distinctive landscapes with the unique cultural essence to themselves.

# 6. Conclusions

There are systematical arguments made for the distinctive insights into contemporary large-scale urban park models on derelict lands through the cross-cultural study between North America, Germany, and China. In addition to demonstrating dynamic landscapes in urban regions in terms of various forms, ideas, structures, approaches, and strategies in urban regeneration across the world, they also put forward the conception of cities as complex mega-systems. There are an increasing number of landscape architects contributing to the exploration into the large-scale urban park evolution in theory and practice, thus shaping "inclusive, safe, resilient and sustainable" cities and human settlement, according to "Transforming Our World: The 2030 Agenda for Sustainable Development".

The theme of expansive parks on post-industrial sites leads us to critically rethink the potential and capability of landscape in urban contexts for the reorganization and redevelopment of the lands, as well as the role and value of landscape architecture as a profession. This should be a lasting belief among landscape architects. As "A Declaration of Concern" declaimed by a group of landscape architects who shared a concern for the quality of the American environment and its future in 1966, a sense of environmental crisis has brought us together. What is merely offensive or disturbing today poses a threat to life itself tomorrow. Currently, there are more intricate challenges and crises arising. However, there is no "single solution" but groups of solutions carefully relating to one another. There is neither a one-shot cure nor a single-purpose panacea, but there is a need for collaborative solutions. The solution to these crises lies in landscape architecture, a profession aimed at the interdependence of environmental processes (LAF 1966).

In this book, North American, German, and Chinese park design paradigms are marked with organic, structural, and shan-shui identities, respectively. James Corner and Peter Latz proposed two park models that reflect the deep cultural integration of North American *cultural imagination* and German cultural contextualization, which is in line with the critical rationalism approaches of *critical thinking* and critical structuralism. Through the analysis conducted in this book, a discovery is made as to the corresponding relationships among critical approach, cultural embedding, the core of shaping the urban landscape, and park identity, as summarized in Table 6.

According to the findings, the critical rationalism method plays a role in guiding and promoting the diverse developments of large-scale urban parks, which improves an intrinsic cultural understanding of urban landscapes shaped by the core of the organism. As the subject of discussion conducted along two remarkable tracks, North American large parks are interpreted as organic infrastructures organized by a dynamic, functioning matrix for the resilient urban landscape. In contrast, German large parks are interpreted as unique urban spaces organized by an open spatial structure for the characteristic urban landscape.

**Table 6.** Relationships among critical rationalism approach, cultural embedding, core of shaping urban landscape and park identity for North America and Germany.

| Category | |
| --- | --- |
| **Germany** | **North America** |
| *Critical structuralism* | *Critical thinking* |
| Cultural contextualization | *Cultural imagination* |
| Difference in the characteristic urban landscape | Organism for the resilient urban landscape |
| Structural identity: as the unique urban space organized by an open spatial structure | Organic identity: as the organic infrastructure organized by a dynamic and functioning matrix |

Source: Table by author.

A reasonable explanation can be made as to the conceptual approaches, theoretical formulations, and representative project cases of the aforementioned chain of relationships for large-scale urban parks. Moreover, it is possible to deduce the significant differences between them to a large extent. In Chapter 5, it is argued that the two urban landscapes and large-scale parks differ.

The critical rationalism approach is crucial to the constant evolution of analysis and understanding of urban landscapes and large parks. Involving persistent rethinking and criticism, the approach makes landscape architecture more professional and critical. Especially, the critical approach is embraced by landscape architecture as a discipline that requires the combination of theory with practice. It prompts landscape architects to review their ideas with caution, discover some parts of falsifiability critically, and provide other alternative insights into the interaction. For the critical readjustments or reconstructions in the urban landscapes in North America and Germany, landscape architects adopted this approach to criticize modern functionalism while advancing the ideas of *landscape urbanism* and *landscape structuralism*.

By taking such an approach, one of the most important parts is observed in regional cultural contexts to explore the way forward for the contemporary urban landscape. In this book, the contributions are reflected in three points of difference in urban landscape between North America and Germany: landscape understanding (coherent vs. creative), landscape and ecology (representation vs. metaphor), and landscape and life (diversity vs. unpredictability). With completely different insights provided into urban landscapes in theories and projects, the development of large-scale urban parks is guided according to the views of Peter Latz and James Corner.

In the meantime, the German structuralistic park design paradigm is aimed at the shaping of open structural spaces with site-specific, characteristic elements through the application of a representation technique that lays emphasis on spatial qualities according to the information in the coherent understanding of landscape. Based on the imagination in the creative landscape understanding, the North

American organic model is intended to highlight spatial performance by building a fluid, adaptive field with non-site conceptual elements through imaging techniques. Within the German park space, the cultivated natural process takes place through the artistic interpretation and representation of nature, which leads to the generation of a diverse urban social life for satisfying the various needs and desires of individuals. In the North American park field, however, the productive natural process is initiated through the metaphor for the ecological agent. Moreover, unpredictable urban social life is stimulated by the flexible, adaptive programs that cause change to social demands and lead to an uncertain future.

As suggested by these results, an answer has been provided to the research question about how the large-scale parks in two developed regions are regarded in terms of contemporary urban social and ecological settings. Arguing for the research hypothesis, they stated that the two large-scale urban park models constructed with their own critical approaches demonstrate the rethinking and conceptions of parks on derelict post-industrial sites.

On this basis, there is a reflection on the Chinese urban landscape and its shan-shui parks. Whether at present or in the future, it is necessary to deeply explore the shan-shui structural method, the role of landscape and related ecological views, and the shan-shui parks in a more critical way according to common points of North American and German large-scale parks as well as the concrete Chinese challenges about landscape conception at the urban level. For the city beautiful conception, they ought to give primary consideration to urban landscapes, not the narrow definition of beautified landscape. With regard to the distinct regional and cultural identity of each park, it should be gradually established by increasing urban practical projects for site renewal and transformation. It would dictate the trend of shan-shui parks. Moreover, it is possible for the trend of unique cultural embedding to arise from the interaction between the ideas and projects of Chinese landscape architects over time.

However, it is unrealistic to form the concrete cultural identity of Chinese shan-shui parks overnight, especially in the contemporary complex and dynamic urban context. Given two relationships with cities and urban nature, the parks have a close connection with Chinese urban spatial structure, particularly when there could be a significant change in the near future to the opposite relationship between the urban and rural areas with the rise of suburban space. It is necessary to examine them from a wider urban and spatial perspective, which is similar to two large-scale urban park conceptions in relation to their revised cities. Allowing for the relationship with urban nature, it is worthwhile to consider developing the shan-shui parks through updated interpretations of contemporary nature in combination with technologies for site reclamation. Furthermore, with the dynamic ecological ideas practiced in the two large-scale parks as eco-machines for processing, there remains a long way to go for Chinese shan-shui parks and ecological viewpoints.

To conclude, the critical, ongoing understanding of contemporary urban landscapes can be improved by the overall research on three design paradigms of large-scale parks on post-industrial sites. Moreover, it is always the case that the regional cultural embedding is linked to the identity of each park. It is expected that the deduction of urban landscapes and large-scale urban parks in the comparative research could provide some rewarding experiences for further research in both China and the rest of the world.

# Appendix A

**Table A1.** North American and German urban landscape analyses as a theoretical foundation.

| Timeline | North American Analysis | German Analysis |
| --- | --- | --- |
| 1970s–1980s | J. B Jackson's analysis: "Landscape Three," a "dynamic system of manmade spaces," impelling North American urban landscape to be process-oriented. | Henri Lefèbvre's analysis: "social production of space," guiding German urban landscape towards the level of "difference". |
| 1990s | "Metaphor" as an orientation: 1. The metaphorical conceptualization of cities through the "lens" of landscape; 2. The ecological metaphor for cities as fluid, living organisms. | "Theoretical construct" as an orientation: For inquiring into new urban spaces, theoretical construct opens up an interdisciplinary path for the analysis and planning of urban regions. |

Source: Author's compilation based on data from Jackson 1984; Lefèbvre [1974] 1991; Corner 2006; Ipsen and Weichler 2005.

**Table A2.** North American and German urban landscape formulations in two theoretical schools of thoughts.

| Timeline | North American Formulation | German Formulation |
| --- | --- | --- |
| 1990s-now | North American *landscape urbanism* with an organic approach. | German *landscape structuralism* with a *structuralistic approach*. |
| | 1. The organic approach embodying the creative potential of ecology in the field of landscape; 2. Employing terms, conceptual categories and operating methodologies of field ecology for understanding sites and cities. | 1. Peter Latz's structuralistic idea highlighting the spatial structure with "informational layers", and their relationships, as an approach to landscape analysis and conception the analysis and planning of urban regions; 2. André Corboz's landscape viewed as "palimpsest". |

Source: Author's compilation based on data from Latz 2008a; Weilacher 2008; Corboz 1983.

**Table A3.** Analytical stage of explaining urban landscapes in Europe and North America.

| Contemporary Urban Landscape | Analytical Stage in Europe | Analytical Stage in North America |
| --- | --- | --- |
| Time | At beginning of the 1970s | During the early 1980s |
| Representative Personality | Henri Lefèbvre | J. B. Jackson |
| Perspective | 1. The ubiquitous globalization and urbanization<br>2. Urban landscape conception in everyday world | |
| Emphasis | "Social production of space," characterized by "difference", "diversity" and "coherence" (Lefèbvre [1974] 1991). | "Vernacular-mobile" landscape, characterized by "dynamic system of manmade space" (Jackson 1984) |

Source: Author's compilation based on data from Jackson 1984; Lefèbvre [1974] 1991.

**Table A4.** Theoretical stages of urban landscapes in Germany and North America.

| Contemporary Urban Landscape | Theoretical Stage in Germany | Theoretical Stage in North America |
| --- | --- | --- |
| Theoretical Orientation | As a "theoretical construct" | As a "metaphor" |
| Theoretical School of Thought | *Landscape structuralism* school of thought | *Landscape urbanism* school of thought |
| Time of Emergence | Since the 1980s | In the mid-1990s |
| Representative Personality | Peter Latz | James Corner |
| Practical Project | Landscape Park Duisburg-Nord; Riemer Park | Freshkills Park; Downsview Park |
| Conceptual Approach | Structuralistic approach | Organic approach |
| Focused View | View of Structuralism in German landscape architecture | Pragmatic and process-oriented North American landscape architecture |

Source: Author's compilation based on data from Corner 2006; Ipsen and Weichler 2005.

**Table A5.** Comparison between German structuralistic parks and North American organic parks from perspectives of Peter Latz and James Corner, based on three-faceted comparison of urban landscapes.

| At the Level of Contemporary Large-Scale Urban Parks | Germany | North America |
| --- | --- | --- |
| **Structuralistic and Organic** | The structural identity | The organic identity |
| **Information and Imagination** | To analyze or decode the physical site and its context, aligning with the coherent landscape understanding in cultural contextualization | An "eidetic scope of landscape creativity" to primarily conceive of sites in *cultural imagination* |
| **Representation and Imaging** | Interaction between objects and subjects, a process of objectification | A mental conception, for improving and creating diverse forms of representational techniques |
| **Spaces and Fields** | In the wake of the spatial turn, shaping urban spaces prepared for social appropriation in everyday life | Establishing fluid adaptive fields, able to absorb, transform, and exchange information with their surroundings |
| **Qualities and Performance** | 1. Qualitatively develop space in all its facets and dimensions 2. Spatial qualities developed by concrete spatial forms of urban life and urban nature for enhancing spatial characteristic | 1. The formative effects of landscape in time 2. Spatial performance produced by dynamic interactions of ecological spatial patterns over time |
| **Cultivated and Productive** | Embracing not merely natural process initiated and sustained by technologies, but also artistic representation of nature by artifacts as a symbol | Producing the effectiveness of nature |

**Table A5.** *Cont.*

| At the Level of Contemporary Large-Scale Urban Parks | Germany | North America |
|---|---|---|
| **Site-specific and Non-site** | Site-specific, meaningful structural elements as existing reference constituting the multilayered open, structure: Topographic, hydrological morphology, water systems, transportation systems, building structures, open space systems, and additional relevant networks | Non-site elements as conceptual diagrams constituting the multilayered, open-ended landscape matrix: Spatial patterns in landscape-ecological sense, such as patch, edge, corridor, and so on, as conceptual elements |
| **Qualitative Characteristics** | See Table 5 | |

Source: Author's compilation based on data from Weilacher 2008; Corner 1999.

# References

Allen, Stan. 1999. *Points + Lines: Diagrams and Projects for the City*. New York: Princeton Architectural Press, pp. 52–53.

Allen, Stan. 2002. Mat Urbanism: The Thick 2-D. In *Case: Le Corbusier's Venice Hospital and the Mat Building Revival*. Edited by Hashim Sarkis, Pablo Allard and Timothy Hyde. Munich: Prestel, pp. 118–26.

Assargård, Hanna. 2011. Landscape Urbanism—From a Methodological Perspective and a Conceptual Framework. Master's thesis, Swedish University of Agricultural Sciences, Uppsala, Sweden; p. 43.

Auer, Sabine. 2010. The Emscher Landscape Park. In *Unter freiem Himmel: Emscher Landschaftspark/Under the Open Sky: Emscher Landscape Park German and English Edition*. Edited by Regionalverband Ruhr. Basel: Birkhäuser, p. 14.

Barrows, Naraelle. 2010. Reinventing Traditionalism: The Influence of Critical Reconstruction on the Shape of Berlin's Friedrichstadt. *Intersections* 11: 55–99.

Bauer, Alexandra, Julian Schaefer, Soeren Schoebel, and Yuting Xie. 2018. Urban Landscape Infiltrations. In *Porous City from Metaphor to Urban Agenda*. Edited by Sophie Wolfrum, Heiner Stengel, Florian Kurbasik, Norbert Kling and Sofia Dona. Basel: Birkhäuser, pp. 226–29.

Beard, Peter. 1996. Peter Latz, Poet of Polltion. *Blueprint* 130: 28–37.

Beck, Ulrich. 2008. Interview with Ulrich Beck: Nationalism Does Not Leave Much Room for the Recognition of Others. Text by Daniel Gamper. *Barcelona Metropolis*. Available online: https://web.archive.org/web/20100701012232/http://www.barcelonametropolis.cat/en/page.asp?id=21&ui=72 (accessed on 1 March 2017).

Beck, Ulrich, and Christoph Lau. 2005. Theorie und Empirie Reflexiver Modernisierung. *Soziale Welt* 3: 107–35. [CrossRef]

Belle, Iris. 2008. Beijing Olympic Forest Park: The Axis to Nature. *Topos* 63: 22–28.

Berger, Alan. 2006. *Drosscape: Wasting Land in Urban America*. New York: Princeton Architectural Press, p. 21.

Berger, Alan. 2009. *Systemic Design can Change the World*. Amsterdam: SUN Publishers.

Bergson, Henri. 1944. *Creative Evolution*. New York: Modern Library, p. 139.

Berrizbeitia, Anita. 2001. Scales of Undecidability. In *Downsview Park Toronto*. Edited by Julia Czerniak. New York: GSD & Prestel, pp. 116–25.

Berrizbeitia, Anita. 2007. Re-placing Process. In *Large Parks*. Edited by Julia Czerniak and George Hargreaves. New York: Princeton Architectural Press, pp. 175–97.

Blackburn, Simon. 2008. *Oxford Dictionary of Philosophy*, 2nd ed. Oxford: Oxford University Press.

Bloomberg, Michael R. 2010. Mayoral Foreword. In *High Performance Landscape Guidelines 21st Century Parks for NYC*. Edited by Charles McKinney, Chelsea Mauldin and Cynthia Gardstein. New York: Design Trust for Public Space and City of New York Parks & Recreation, p. 6.

BMU. 2019. *Masterplan Stadtnatur Maßnahmenprogramm der Bundesregierung für eine Lebendige Stadt*. Berlin: BMU.

Böhmethus, Hartmut. 2000. Wer Sagt, Was Leben Ist? Available online: http://www.zeit.de/2000/49/200049_g-boehme.xml (accessed on 1 May 2017).

Bruegmann, Robert. 2008. Broadacre City and Sprawls. In *Multiple City: Stadtkonzepte 1908 bis 2008*. Edited by Sophie Wolfrum and Winfried Nerdinger. Berlin: Jovis, pp. 54–57.

Bucher, Annemarie. 2013. Landscape Thoeries in Transition Shifting Realities and Multiperspectives Perception. In *Topology Landscript 3*. Edited by Christophe Girot, Anette Freytag, Albert Kirchengast and Dunja Richter. Berlin: Jovis, pp. 35–48.

Burckhardt, Lucius, and Bazon Brock. 1985. *Die Kinder fressen ihr Revolution. Wohnen-Planen-Bauen-Grünen*. Cologne: DuMont, p. 241.

Capra, Fritjof. 1996. *The Web of Life: A New Scientific Understanding of Living Systems*. New York: Anchor Books.

Chen, Xiangqiang, and Jianguo Wu. 2009. Sustainable Landscape Architecture: Implications of the Chinese Philosophy of "Unity of Man with Nature" and Beyond. *Landscape Ecology* 8: 1015.

Choay, Françoise. 1985. Critique. *Princeton Journal of Architecture* 2: 211–20.

Clemmensen, Thomas Juel. 2015. The Garden and The Machine. In *Revising Green Infrastructure Concepts Between Nature and Design*. Edited by Daniel Czechowski, Thomas Hauck and Georg Hausladen. Boca Raton: CRC Press, Taylor & Francis Group, pp. 137–52.

Cohen, Paul E. 1997. *Manhattan in Maps: 1527–1995*. New York: Rizzoli, pp. 100–05.

Corboz, André. 1983. The Land as Palimpsest. *Diogenes* 121: 12–34. [CrossRef]

Corner, James. 1991. Critical Thinking and Landscape Architecture. *Landscape Journal* 2: 115–33. [CrossRef]

Corner, James. 1997. Ecology and landscape as agents of creativity. In *Ecological Design and Planning*, 1st ed. Edited by George F. Thompson and Frederick R. Steiner. New York: John Wiley and Sons, INC, pp. 81–108.

Corner, James. 1999. *Recovering Landscape: Essays in Contemporary Landscape Architecture*. New York: Princeton Architectural Press.

Corner, James. 2005. Lifescape—Freshkills Parkland. *Topos* 51: 14–21.

Corner, James. 2006. Terra Fluxus. In *The Landscape Urbanism*. Edited by Charles Waldheim. New York: Reader Princeton Architectural Press, pp. 21–33.

Corner, James. 2007. Foreword. In *Large Parks*. Edited by Julia Czerniak and George Hargreaves. New York: Princeton Architectural Press, pp. 11–14.

Corner, James. 2009. Shelby Farms Park Strategies for a Large Urban Park in Memphis, USA. *Topos* 66: 16–20.

Corner, James. 2021. Plot. In *250 Things a Landscape Architect Should Know*, 1st ed. Edited by B. Cannon Ivers. Basel: Birkhäuser.

Corner, James, and Alison Bick Hirsch. 2014. *The Landscape Imagination Collected Essays of James Corner 1990–2010*. New York: Princeton Architectural Press.

Corner, James, and Stan Allen. 2001. Emergent Ecologies. In *Downsview Park Toronto*. Edited by Julia Czerniak. New York: GSD & Prestel, pp. 58–65.

Cosgrove, Denis. 1985. Prospect, Perspective and the Evolution of the Landscape Idea. *Transactions of the Institute of British Geographers* 10: 45–62. [CrossRef]

Cosgrove, Denis, and Stephen Daniels. 1988. *The Iconography of Landscape: Essays on the Symbolic Representation, Design and Use of Past Environments*. New York: Cambridge University Press.

Cranz, Galen. 1982. *The Politics of Park Design: A Hitory of Urban Parks in America*. Cambridge: The MIT Press.

Culler, Jonathan. 2002. *Structuralist Poetics: Structuralism, Linguistics and The Study of Literature*, 2nd ed. London and New York: Routledge.

Czechowski, Daniel, Thomas Hauck, and Georg Hausladen. 2015. *Revising Green Infrastructure: Concepts Between Nature and Design*. Boca Raton: CRC Press, Taylor & Francis Group.

Czerniak, Julia. 2001. *Downsview Park Toronto*. New York: Prestel Pub.

Czerniak, Julia. 2022. Large Parks: Trends (and Possibilities). In *Why Cities Need Large Parks Large Parks in Large Cities*. Edited by Richard Murray. London and New York: Routledge Taylor and Francis Group, pp. 412–27.

Czerniak, Julia, and George Hargreaves. 2007. *Large Parks*. New York: Princeton Architectural Press.

De Jong, E. A. 2000. Paradise is just where you are right now. In *Open Spaces*. Edited by Günther Vogt. Basel,·Boston and·Berlin: Birkhäuser Verlag für Architektur, pp. 10–15.

Deleuze, Gilles. 2002. *Desert Islands and Other Texts 1953–1974*. New York: Semiotexte, p. 170.

Deleuze, Gilles, and Félix Guattari. 2004. *A Thousand Plateaus: Capitalism and Schizophrenia*. London: Continuum International Publishing Group Ltd.

Derrida, Jacques. 1967. *Of Grammatology Corrected Edition*. Translated by Gayatri Chakravorty Spivak. Baltimore and London: Johns Hopkins University Press.

Dettmar, Jörg, and Karl Ganser. 1999. *Industrie Natur Ökologie und Gartenkunst im Emscher Park*. Stuttgart,·Berlin·and Köln: Verlag Eugen Ulmer, pp. 134–53.

Dettmar, Jörg, and Udo Weilacher. 2003. Landscape as a Process. *Topos* 44: 76–81.

Dümpelmann, Sonja. 2018. A New Paradigm for the Promenade. In *Inspiration Highline*. Edited by Udo Weilacher. Munich: Technical University of Munich, pp. 24–25.

Duncan, Allison, and Ethan Seltzer. 2010. Landscape Urbanism: An Annotated Bibliography. Available online: http://www.terrafluxus.com/wp-content/uploads/2010/10/final-format-LU-bib-2.pdf (accessed on 15 June 2022).

Eisel, Ulrich. 1982. Die schöne Landschaft als kritische Utopie oder als konservatives Relikt. *Soziale Welt* 2: 157–68.

Fabris, Luca M. F. 1999. IBA Emscher Park 1989–1999 and Beyond. *Abitare* 386: 99–115.

Fabris, Luca M. F., and Mengyixin Li. 2020. 历史性视角下后工业区景观认知及实践变革研究 Research on Post-industrial Area Landscape Cognition and Practice Transformation From a Historical Perspective. 风景园林 *Landscape Architecture* 7: 8–17.

Field Operations. 2001. Lifescape Freshkills Landsfill to Landscape Design Competition Staten Island, New York. Available online: http://www1.nyc.gov/assets/planning/download/pdf/plans/fkl/fien1.pdf (accessed on 1 July 2022).

Field Operations. 2006. Freshkills Park: Lifescape Draft Master Plan. Available online: http://www.nyc.gov/html/dcp/html/fkl/fkl_index.shtml (accessed on 1 July 2022).

Forman, Richard T. T. 1987. The Ethics of Isolation, the Spread of Disturbance, and Landscape Ecology. In *Landscape Heterogeneity and Disturbance*. Edited by Monica Goigel Turner. New York: Springer.

Forman, Richard T. T. 2022. Values of Large-versus-Small Urban Greenspaces and their Arrangement. In *Why Cities Need Large Parks Large Parks in Large Cities*. Edited by Richard Murray. London and New York: Routledge Taylor and Francis Group, pp. 18–43.

Forman, Richard T. T., and Michel Godron. 1986. *Landscape Ecology*. New York: John Wiley & Sons.

Forman, Richard T. T., Wenche E. Dramstad, and James D. Olson. 1996. *Landscape Ecology Principals in Landscape Architecture and Land-Use Planning*. Washington, DC: GSD, American Society of Landscape Architects, and Island Press, p. 32.

Fulton, Gale. 2003. Large Parks: New Perspectives. A Symposium at the Harvard Graduate School of Design, Cambridge, Massachusetts, 10–12 April 2003. *Landscape Journal* 2: 171–73.

Fung, Stanislaus, and Hongde Wu. 2020. Between Sapience and Sentience: Four Remarks in Response to Beautiful China. In *Beautiful China Reflections on Landscape Architecture in Contemporary China*. Edited by Richard J. Weller and Tatum L. Hands. Los Angeles, San Francisco, New York and Shenzhen: ORO Editions, pp. 198–207.

Gailing, Ludger. 2005. Sustainable Landscape Development with Regional Parks. Overcoming Problems of Landscape Multifunctionality in Urban Agglomerations. Paper presented at ERSA Congress Land Use and Water Management in a Sustainable Network Society, Amsterdam, The Netherlands, August 23–27.

Ganser, Karl. 1991. Die Strategie der IBA Emscher Park. *Garten + Landschaft* 10: 15.

Ganser, Karl, Siebel Walter, and Thomas Sieverts. 1993. Die Planungsstrategie der IBA Emscherpark. *Raumplaung* 61: 112–18.

Geddes, Patrick. 1915. *Cities in Evolution*. London: Williams.

Giseke, Undine. 2018. The City in the Anthropocene—Multiple Porosities. In *Porous City from Metaphor to Urban Agenda*. Edited by Sophie Wolfrum, Heiner Stengel, Florian Kurbasik, Norbert Kling and Sofia Dona. Basel: Birkhäuser, pp. 200–3.

Glover, Robert. 2001. City Making and the Making of Downsview Park. In *Downsview Park Toronto*. Edited by Julia Czerniak. New York: GSD & Prestel, pp. 34–39.

Godau, Sigrid, and Claudia Heinrich. 2010. Landschaftspark Duisburg-Nord. In *Unter freiem Himmel: Emscher Landschaftspark/Under the Open Sky: Emscher Landscape Park German and English Edition*. Edited by Regionalverband Ruhr. Basel: Birkhäuser, pp. 64–71.

Gray, Christopher D. 2006. From Emergence to Divergence: Modes of Landscape Urbanism. Dissertation thesis, Edinburgh College of Art School of Architecture, Edinburgh, UK.

Grosch, Leonard, and Constanze A. Petrow. 2021. *Designing Parks Berlin's Park am Gleisdreieck or the Art of Creating Lively Places*. Berlin: Jovis.

Haber, Wolfgang. 2010. Post-Industrial Cultural Landscapes. In *Field Studies The New Aesthetics of Urban Agriculture*. Edited by Regionalverband Ruhr and Udo Weilacher. Basel: Birkhäuser, pp. 16–27.

Haberl, Helmut, Dominik Wiedenhofer, and Stefan Pauliuk. 2019. Contributions of Sociometabolic Research to Sustainability Science. *Nature Sustainability* 2: 173–84. [CrossRef]

Hauck, Thomas, and Daniel Czechowski. 2015. Green Functionalisim a Brief Sketch of Its History and Ideas in the United States and Germany. In *Revising Green Infrastructure Concepts Between Nature and Design*. Edited by Daniel Czechowski, Thomas Hauck and Georg Hausladen. Boca Raton: CRC Press, Taylor & Francis Group, pp. 3–28.

Hertzberger, Herman. 1981. The Archaic Principles of Human Nature Synchronous Thinking in Structuralism. In *Structuralism in Architecture and Urban Planning*. Edited by Arnulf Lüchinger. Stuttgart: Krämer Verlag, p. 24.

Hill, Kristina. 2001. Urban Ecologies: Biodiversity and Urban Design. In *Downsview Park Toronto*. Edited by Julia Czerniak. New York: GSD & Prestel, pp. 90–101.

Hines, Thomas S. 1988. No Little Plans: The Achievement of Daniel Burnham. *Museum Studies* 13: 105. [CrossRef]

Höfer, Wolfram. 2013. Landschaftsurbanismus. In *StadtGrün*. Edited by Jirku Almut. Stuttgart: Fraunhofer IRB Verlag, pp. 74–80.

Höfer, Wolfram, and Ludwig Trepl. 2010. Jackson's Concluding with Landscapes—Full Circle. *Journal of Landscape Architecture* 5: 40–51. [CrossRef]

Höfer, Wolfram, and Vera Vicenzotti. 2013. Post-industrial Landscapes: Evolving Concepts. In *The Routledge Companion to Landscape Studies*, 1st ed. Edited by Peter Howard, Ian Thompson, Emma Waterton and Mick Atha. London and New York: Routledge Taylor and Francis Group, pp. 405–16.

Hoffmann, Lutz. 2008. City/Building/Plan: 850 Years of Urban Development in Munich. Landeshauptstadt München, Referat für Stadtplanung und Bauordnung. Available online: https://stadt.muenchen.de/infos/exhibition-city-building-plan.html (accessed on 1 March 2014).

Holling, C. S. 2001. Understanding the Complexity of Economic, Ecological, and Social Systems. *Ecosystems* 4: 390–405. [CrossRef]

Hu, Jie. 2020. 山水城市 梦想人居—基于山水城市思想的风景园林规划设计实践 *Shan-Shui City Ideal Settlements: Landscape Planning and Design Practice Based on the Ideology of Shan-Shui City*. Beijing: China Architecture & Building Press.

Huyssen, Andreas. 1986. *After the Great Divide: Modernism, Mass Culture, Postmodernism.* Bloomington: Indiana University Press.

Ipsen, Detlev, and Holger Weichler. 2005. Landscape Urbanism. *Monu-Magazine on Urbanism, Middle Class Urbanism,* 39–47.

Jackson, John Brinckerhoff. 1984. *Discovering the vernacular landscape.* New Haven and London: Yale University Press.

Jacobs, Jane. 1961. *The Death and Life of Great American Cities.* New York: Vintage Books.

Jencks, Charles. 1977. *The Language of Post-Modern Architecture.* New York: Rizzoli.

Kirchhoff, Thomas, and Ludwig Trepl. 2009. *Vieldeutige Natur. Landschaft, Wildnis und Ökosystem als Kulturgeschichtliche Phänomene.* Bielefeld: Transcript, p. 25.

Kirkwood, Niall. 2004. The Anti-Olmsted by Arthur Lubow. *The New York Times Magazine,* 46–52.

Kleihues, Josef Paul, and Christina Rathgeber. 1993. *Berilin-New York: Like and Unlike: Essays on Architecture and Art from 1870 to the Present.* New York: Rizzoli International Publications.

Kolkau, Anette. 2002. Emscher Landscape Park in the Post-IBA Era. *Topos* 40: 32–38.

Koolhaas, Rem, and Bruce Mau. 1995. *S, M, L, XL. O.M.A. Rem Koolhaas and Bruce Mau.* New York: The Monacelli Press.

Körner, Stefan. 2013. Landscape and Modernity. In *Topology Landscript 3.* Edited by Christophe Girot, Anette Freytag, Albert Kirchengast and Dunja Richter. Berlin: Jovis, pp. 117–36.

Kowarik, Ingo. 1992. Das Besondere der städtischen Flora und Vegetation. *Schriftenreihe des Deutschen Rates für Landespflege* 61: 33–47.

Küchler-Krischu, Jonna, Christiane Schell, Karl-Heinz Erdmann, and Andreas W. Mues. 2016. *2015 Nature Awareness Study Population Survey on Nature and Biological Diversity;* Berlin: Federal Ministry for the Environment, Nature Conservation, Building and Nuclear Safety.

Kühn, Norbert. 2013. The Relationship Between Plants and Landscape Architectural Design—An Attempt at Repositioning. In *Topology Landscript 3.* Edited by Christophe Girot, Anette Freytag, Albert Kirchengast and Dunja Richter. Berlin: Jovis, pp. 139–60.

Küster, Hansjörg. 2013. Landscapology Linking Natural Sciences, Humanities, and Aesthetics. In *Topology Landscript 3.* Edited by Christophe Girot, Anette Freytag, Albert Kirchengast and Dunja Richter. Berlin: Jovis, pp. 163–81.

LAF. 1966. A Declaration of Concern. Available online: https://www.lafoundation.org/who-we-are/values/declaration-of-concern (accessed on 20 May 2022).

Lampugnani, Vittorio. 1983. The Facts and the Dreams: AD Interview. *Architectural Design* 53: 17–19.

Latz, Peter. 1993. Design by Handling the Existing. In *Modern Park Design: Recent Trends.* Edited by Martin Knit, Hans Opus and Peter van Sane. Amsterdam: Thoth, pp. 92–97.

Latz, Peter. 2003. The Idea of Making Time Visible. *Topos* 33: 77–82.

Latz, Peter. 2004. Landscape Park Duisburg-Nord: The Metamorphosis of an Industrial Site. In *Manufactured Sites: Rethinking the Post-Industrial Landscape.* Edited by Niall Kirkwood. London and New York: Routledge Taylor and Francis Group, pp. 149–61.

Latz, Peter. 2005. Landscape architecture as an Intercultural Principle. *Topos* 50: 6–12.

Latz, Peter. 2008a. Design is Experimental Invention. In *Creating Knowledge Innovation Strategies for Designing Urban Landscapes.* Edited by Hille von Seggern, Julia Werner and Lucia Grosse-Bächle. Berlin: Jovis, pp. 332–61.

Latz, Peter. 2008b. Duisburg Nord Landscape Park. In *Syntax of Landscape: The Landscape Architecture of Peter Latz and Partners.* Edited by Udo Weilacher. Basel, Boston and Berlin: Birkhäuser, pp. 102–33.

Latz, Peter. 2008c. Vision und Aktion. *Garten + Landschaft* 3: 8–9.

Latz, Peter. 2012. Der Park des 21. Jahrhunderts. Available online: https://www.emeriti-of-excellence.tum.de/fileadmin/w00bpl/www/Veranstaltungsarchiv/Vortraege_Highlights-der-Forschung/2012-12-13_Latz_Parks_Zusammenfassung.pdf (accessed on 15 June 2023).

Latz, Peter. 2013a. Landscape Park Duisburg-Nord—Chaos Remains Chaos, Five Layers of a Transformation. *Topos* 84: 104–07.

Latz, Peter. 2013b. Parco Dora Turin—Transformation of an industrial brownfield site. *Topos* 84: 102–03.

Latz, Peter. 2015. Peter Latz: Rehabilitating Postindustrial Landscapes by Stephen Heyman in New York Times. Available online: https://www.nytimes.com/2015/12/31/arts/international/peter-latz-rehabilitating-postindustrial-landscapes.html (accessed on 10 May 2016).

Latz, Peter. 2016a. Informationsdichte von Landschaft im Gespräch mit Peter Latz. In *Land-Schafts Vertrag zur Kritischen Rekonstruktion der Kulturlandschaft*. Edited by Sören Schöbel. Berlin: Jovis, pp. 155–63.

Latz, Peter. 2016b. The Park in the 21st Century. In *Rust Red Landscape Park Duisburg-Nord*. Munich: Hirmer, pp. 262–67.

Lefèbvre, Henri. 1991. *The Production of Space*. Translated by Donald Nicholson-Smith. Oxford: John Wiley & Sons, p. 64. First Published 1974.

Lefèbvre, Henri. 2003. *The Urban Revolution*, 1st ed. Translated by Robert Bononno. Minneapolis and London: University of Minnesota Press. First Published 1970.

Lewis, James R., and Evelyn Dorothy Oliver. 2009. *The Dream Encyclopedia*. Detroit: Visible Ink Press, p. 204.

Li, Feng, Rusong Wang, Juergen Paulussen, and Xusheng Liu. 2005a. Comprehensive Concept Planning of Urban Greening Based on Ecological Principles: A Case Study in Beijing, China. *Landscape and Urban Planning* 4: 325–36. [CrossRef]

Li, Mengyixin. 2022a. Landscape Regeneration of Urban Industrial Heritage Sites Towards the Human Factors Perspectives: Perception and Assessment. *Landscape Architecture and Regional Planning* 1: 1–7. [CrossRef]

Li, Mengyixin. 2022b. 德国城市自然整体规划研究与启示 Research on and Enlightenment of the Overall Planning of Urban Nature in Germany. 风景园林 *Landscape Architecture* 6: 70–75.

Li, Mengyixin. 2023. 批判性城市景观重构下的美国与德国后工业景观比较 Comparison of Post-Industrial Landscape Between the USA and Germany in the Context of Critical Reconstruction of Urban Landscape. 风景园林 *Landscape Architecture* 6: 12–19.

Li, Weifeng, Zhiyun Ouyang, and Rusong Wang. 2005b. Land Potential Evaluation for Large-scale Greenbelt Development at Urban-rural Transition Zone A Case Study of Beijing, China. *Remote Sensing for Land & Resources* 1: 53–56.

Lister, Nina-Marie. 2007. Sustainable Large Parks: Ecological Design or Designer Ecology? In *Large Parks*. Edited by Julia Czerniak and George Hargreaves. New York: Princeton Architectural Press, pp. 31–51.

Lister, Nina-Marie. 2015. Resilience Designing the New Sustainability. *Topos* 90: 14–20.

Lister, Nina-Marie. 2016. Interview with Nina-Marie Lister by Jared Green. Available online: https://www.asla.org/ContentDetail.aspx?id=31738 (accessed on 1 July 2017).

LMRSB. 1995. *Messestadt Riem, Ideen- und Realisierungswettbewerb Landschaftspark München Riem*. München: Landeshauptstadt München Planungsreferat.

LMRSB. 1998. *Messestadt Riem Ökologische Bausteine Teil II Gebäude und Freiraum*. München: Landeshauptstadt München Planungsreferat, p. 4.

LMRSB. 2005. *Evaluierung Messestadt Riem Nachhaltige Stadtenwicklung in München*. München: Landeshauptstadt München Planungsreferat, p. 16.

LMRSB. 2009. *Messestadt Riem*. München: Landeshauptstadt München Planungsreferat.

Lü, Hui, and Yufan Zhu. 2020. The Imagination in Post-Modernism–Research of Post-industrial Landscape Design in the Case of Qunminghu Park in Shougang Group. *Chinese Landscape Architecture* 3: 27–32.

Lubow, Arthur. 2004. The Anti-Olmsted. *The New York Times Magazine*, May 16, 46–52.

Lüchinger, Arnulf. 1981. *Strukturalism in Architecture and Urban Planning*. Stuttgart: Karl Krämer Verlag.

Ma, Hongyu. 2013a. *American Culture: An Introduction*. Dalian: Dalian Maritime University Press.

Ma, Yansong. 2013b. Ma Yansong's "Shanshui City" Book Launch and Exhibition Held in Beijing. Available online: http://www.i-mad.com/press/ma-yansongs-shanshui-city-book-launch-and-exhibition-held-in-beijing/ (accessed on 1 January 2017).

Marcinkoski, Christopher. 2020. Considering China's Future Public Realm. In *Beautiful China Reflections on Landscape Architecture in Contemporary China*. Edited by Richard J. Weller and Tatum L. Hands. Los Angeles, San Francisco, New York and Shenzhen: ORO Editions, pp. 144–51.

Martin, Geoffrey, Eileen W James, and Preston E. James. 1981. *All Possible Worlds: A History of Geographical Ideas*. New York: John Wiley & Sons, p. 177.

Marton, Deborah. 2010. Design Trust Preface. In *High Performance Landscape Guidelines 21st Century Parks For NYC*. Edited by Charles McKinney, Chelsea Mauldin and Cynthia Gardstein. New York: Design Trust for Public Space and City of New York Parks & Recreation, p. 7.

Marx, Leo. 1964. *The Machine in the Garden*. New York: Oxford University Press, p. 23.

Mau, Bruce. 2000. *Life Style*. London: Phaidon Press, p. 288.

McAvin, Margaret. 1991. Landscape Architecture and Critical Inquiry—Introduction. *Landscape Journal* 2: 155–56. [CrossRef]

McHarg, Ian L. 1969. *Design with Nature*. New York: The Natural History Press.

Mertins, Detlef. 2001. Downsview Park International Design Competition. In *Downsview Park Toronto*. Edited by Julia Czerniak. New York: GSD & Prestel, pp. 24–33.

Meyer, Elizabeth K. 1997. The expanded field of landscape architecture. In *Ecological Design and Planning*, 1st ed. Edited by George F. Thompson and Frederick R. Steiner. New York: John Wiley & Sons, Inc., pp. 45–79.

Meyer, Elizabeth K. 2007. Uncertain Parks: Disturbed sites, Citizens, and Risk Society. In *Large Parks*. Edited by Julia Czerniak and George Hargreaves. New York: Princeton Architectural Press, pp. 58–85.

Meyer, Elizabeth K. 2008. Sustaining Beauty. The Performance of Appearance: A Manifesto in Three Parts. *Journal of Landscape Architecture* 1: 6–23. [CrossRef]

Mitchell, W. J. T. 2002. *Landscape and Power Second Edition*. Chicago and London: The University of Chicago Press.

Mostafavi, Mohsen, and Ciro Najle. 2003. *Landscape Urbanism: A Manual for the Machinic Landscape*. London: Architectural Association.

Murray, Richard. 2022. *Why Cities Need Large Parks Large Parks in Large Cities*. London and New York: Routledge Taylor and Francis Group.

North, Alissa. 2012. Processing Downsview Park: Transforming a Theoretical Diagram to Master Plan and Construction Reality. *Journal of Landscape Architecture* 1: 8–19. [CrossRef]

NYCDPR. 2006. Fresh Kills Park Project. Available online: https://www.nyc.gov/freshkillspark (accessed on 15 May 2023).

NYCDPR. 2021. Freshkills Park: Site History. Available online: https://www.nycgovparks. org/park-features/freshkills-park/about-the-site (accessed on 4 June 2022).

Olmsted, Frederick Law, Jr., and Theodora Kimball. 1928. *Frederick Law Olmsted Landscape Architect 1822–1903*. New York: G. P. Putnam's Sons, p. 27.

Peisl, Julius. 2014. Strukturalismus in der Landschaftsarchitektur Eine Theoretische Untersuchung am Beispiel Peter Latz. Master's thesis, . Technical University of Munich, Munich, Germany.

Pollak, Linda. 2001. Building City Landscape: Interdisciplinary Design Work in the Downsview Park Competition. In *Downsview Park Toronto*. Edited by Julia Czerniak. New York: GSD & Prestel, pp. 40–47.

Pollak, Linda. 2007. Matrix Landscape: Construction of Identity in the Large Park. In *Large Parks*. Edited by Julia Czerniak and George Hargreaves. New York: Princeton Architectural Press, pp. 87–119.

Popper, Karl. 1957. *The Poverty of Historicism*. London: Routledge.

Popper, Karl. 1959. *The Logic of Scientific Discovery*. New York: Harper and Row.

Popper, Karl. 1976. *Unended Quest: An Intellectual Autobiography*. London and New York: Routledge, p. 41.

Poser, Hans. 2008. Creativity in the Balance Between Action and Complexity. In *Creating Knowledge Innovation Strategies for Designing Urban Landscapes*. Edited by Hille von Seggern, Julia Werner and Lucia Grosse-Bächle. Berlin: Jovis, pp. 108–23.

Prominski, Martin. 2005. Designing Landscapes as Evolutionary Systems. *The Design Journal* 3: 25–34. [CrossRef]

Prominski, Martin. 2010. The Landscape Dilemma and its Potential for Optimism in Landscape Architecture. In *Landscape 21 International Journal for Planning Research and Landscape Design*. Slovenia: Department of Landscape Architecture Biotechnical Faculty, University of Ljubljana, pp. 57–62.

Qian, Xuesen. 1996. Letter to Wu Liangyong on the Subject of "Shan-Shui City". In *Qian Xuesen's Theory on Urbanology and Shan-Shui City*. Edited by Shixing Bao. Beijing: China Architecture & Building Press, p. 47.

Rahmann, Heike, and Jillian Walliss. 2020. *The Big Asian Book of Landscape Architecture*. Berlin: Jovis.

Reed, Chris, and Nina-Marie Lister. 2014. *Projective Ecologies*. New York: GSD and ACTAR.

Riehl, W. Heinrich. 1851. *Die Naturgeschichte des Volkes als Grundlage einer deutschen Social Politik. Erster Band. Land und Leute*. Stuttgart: J. G. Cottascher Verlag.

Rohlf, Michael. 2008. The Transition from Nature to Freedom in Kant's Third Critique. *Kant Studien* 3: 339–60. [CrossRef]

Roncken, Paul A., Sven Stremke, and Maurice Paulissen. 2011. Landscape Machines: Productive Nature and the Future Sublime. *Journal of Landscape Architecture* 1: 68–81. [CrossRef]

Rosenberg, Elissa. 2007. Gardens, Landscape, Nature: Duisburg-Nord, Germany. In *The Hand and the Soul-Aesthetics and Ethics in Architecture and Art*. Edited by Sanda Iliescu. Charlottesville: University of Virginia Press, pp. 209–30.

Rossi, Aldo. 1984. *The Architecture of the City*. New York: The MIT Press.

Rossmann, Andreas. 2009. Looking back: IBA Emscher Park. In *Landscape as a System Contemporary German Landscape Architecture*. Edited by Bund Deutscher Landschaftsarchitekten. Basel: Birkhäuser, pp. 148–61.

Rybczynski, Witold. 1995. *City Life: Urban Expectations in A New York*. New York: Scribner.

Salt, George W. 1979. A Comment on the Use of the Term Emergent Properties. *The American Naturalist* 1: 145–48. [CrossRef]

Schäfer, Robert. 2005. With time. *Topos* 91: 7–10.

Schegk, Ingrid, and Sabrina Wilk. 2007. Landscape architecture in Germany Case Study the Landscape Park in Riem. In *European Landscape Architecture: Best practice in detailing*. Edited by Ian Thompson, Torben Dam and Jens Balsby Nielsen. London and New York: Routledge Taylor and Francis Group, pp. 81–106.

Schöbel, Sören. 2006. Qualitative Research as a Perspective for Urban Open Space Planning. *Journal of Landscape Architecture* 1: 38–47. [CrossRef]

Schöbel, Sören. 2007. Landschaft als Prinzip: Über das Verstehen, Erklären und Entwerfen. *Stadt + Grün* 56: 53–58.

Schöbel, Sören. 2014. Landschaft—Kritische Rekonstruktion. In *Zukunft aus Landschaft Gestalten Stichworte zur Landschaftsarchitektur*. Edited by Hubertus Fischer. München: AVM, pp. 147–52.

Schöbel, Sören. 2015. Landschaft als Strukturgeber an den Rändern der Städte. Über Trenngrün und kritische Rekonstruktion. In *An den Rändern der Städte. Strategien für die Inwertsetzung von Inneren und Äußeren Landschaften in Brandenburg*. Cottbus: Förderverein der Brandenburgischen Technischen Universität Cottbus, pp. 20–25.

Schöbel, Sören. 2018. Im Überblick Gesellschaftsvertrag—Stadtvertrag—Transformation. In *Land—Schafts Vertrag Zur Kritischen Rekonstruktion Der Kulturlandschaft*. Berlin: Jovis, pp. 70–73.

Schöbel, Sören, and Daniel Czechowski. 2009. Rescaling Landscape Architecture. Paper presented at ECLAS Conference 2009 Landscape and Ruins, Genova, Italy, September 23–26.

Schöbel, Sören, and Daniel Czechowski. 2013. *Urban Landscape Studies. Euphorigenic Landscapes—issue 1.0*. Freising: Fachgebiet für Landschaftsarchitektur regionaler Freiräume Technische Universität München.

Schöbel, Sören, and Daniel Czechowski. 2015. Beyond Infrastructure and Superstructure Intermediating Landscapes. In *Revising Green Infrastructure Concepts Between Nature and Design*. Edited by Daniel Czechowski, Thomas Hauck and Georg Hausladen. Boca Raton, London and New York: CRC Press, Taylor & Francis Group, pp. 171–91.

Schöbel, Sören, Andreas R. Dittrich, and Daniel Czechowski. 2013. Energy Landscape Visualization: Scientific Quality and Social Responsibility of a Powerful Tool. In *Sustainable Energy Landscapes Designing, Planning, and Development*. Edited by Sven Stremke and Andy van den Dobbelsteen. Boca Raton, London and New York: CRC Press Taylor & Francis Group, pp. 133–59.

Schumacher, Patrik, and Christian Rogner. 2001. After Ford. In *Stalking Detroit*. Edited by Daskalakis Georgia, Charles Waldheim and Jason Young. Barcelona: Actar, pp. 48–56.

Secchi, Bernardo. 2007. Rethinking and Redesigning the Urban Landscape. *Places* 19: 6–11.

Selugga, Malte. 2008. The Dragon's Tail. *Topos* 63: 14–21.

Shane, David Grahame. 2006. The Emergence of Landscape Urbanism. In *The Landscape Urbanism Reader*. Edited by Charles Waldheim. New York: Princeton Architectural Press, pp. 55–67.

Shane, David Grahame. 2011. *Urban Design Since 1945—A Global Perspective*, 1st ed. Chichester: John Wiley & Sons Ltd., p. 346.

Shannon, Kelly. 2012. (R)evolutionary Ecological Infrastructures. In *Designed Ecologies: The Landscape Architecture of Kongjian Yu*. Edited by William S. Saunders. Basel: Birkhäuser, pp. 200–11.

Siemer, Stefan, and Ulrike Stottrop. 2010. Castellans, Steel Barons and Leisure Kings: Parks in Cultural History of the Ruhr Area. In *Under the Open Sky Emscher Landscape Park*. Edited by Sabine Auer, Sigrid Godau and Regionalverband Ruhr. Basel: Birkhäuser Verlag GmbH, pp. 52–59.

Sieverts, Thomas. 2003. *Cities without Cities: An interpretation of the Zwischenstadt*. London and New York: Spon Press Taylor & Francis Group.

Sieverts, Thomas. 2008. Improving the Quality of Fragmented Urban Landscapes—A Global Challenge! In *Creating Knowledge. Innovation Strategies for Designing Urban Landscapes*. Edited by Hille von Seggern, Julia Werner and Lucia Grosse-Bächle. Berlin: Jovis, pp. 252–65.

Simmel, Georg. 2007. The Philosophy of Landscape. *Theory, Culture & Society* 24: 22–29.

Smith, Neil. 2003. Foreword. In *The Urban Revolution*, 1st ed. Edited by Henri Lefebvre. Translated by Robert Bononno. Minneapolis and London: University of Minnesota Press, pp. xx–xxi. First published 1970.

Smithson, Robert. 1996. A Sedimentation of the Mind: Earth Projects. In *Robert Smithson: The Collected Writings*, 1st ed. Edited by Jack Flam. Berkeley and Los Angeles: University of California Press, p. 105.

Somol, Robert. 2001. All Systems Go! The Terminal Nature of Contemporary Urbanism. In *Downsview Park Toronto*. Edited by Julia Czerniak. New York: GSD & Prestel, p. 131.

Speaks, Michael. 2006. Intelligence After Theory. In *Perspecta 38 Architecture After All: The Yale Architectural Journal*. Edited by Marcus Carter, Christopher Marcinkoski and Forth Bagley. Cambridge: The MIT Press, pp. 103–6.

Stilgenbauer, Judith. 2005. Landschaftspark Duisburg Nord—Duisburg 2005 Germany EDRA/Places Award Design. *Places* 17: 6–9.

Stokman, Antje, Sabine Rabe, and Stefanie Ruff. 2008. Beijing's New Urban Countryside—Designing with Complexity and Strategic Landscape Planning. *Journal of Landscape Architecture* 2: 30–45. [CrossRef]

Studer, Meg. 2011. An Interview with Charles Waldheim: Landscape Urbanism Now. Available online: https://scenariojournal.com/article/an-interview-with-charles-waldheim/ (accessed on 1 January 2017).

Swaffield, Simon. 2002. *Theory in Landscape Architecture: A Reader*. Philadelphia: University of Pennsylvania Press.

Swaffield, Simon. 2006. Theory and Critique in Landscape Architecture: Making Connections. *Journal of Landscape Architecture* 1: 22–29. [CrossRef]

Tate, Alan. 2004. *Great City Parks*. London and New York: Taylor & Francis.

Thierstein, Alain. 2018. The Connected and Multisalar City: Porosity in the Twenty-first Century. In *Porous City from Metaphor to Urban Agenda*. Edited by Sophie Wolfrum, Heiner Stengel, Florian Kurbasik, Norbert Kling and Sofia Dona. Basel: Birkhäuser, pp. 222–25.

Thompson, Ian. 2012. Ten Tenets and Six Questions for Landscape Urbanism. *Landscape Research* 37: 7–26. [CrossRef]

Treib, Marc. 2009. Field. In *Learning from Duisburg Nord: Comments of International Experts on a Masterpiece of Contemporary Landscape Architecture*. Edited by Udo Weilacher. Müchen: Technische Universität München, Lehrstuhl für Landschaftsarchitektur und Industrielle Landschaft, p. 66.

Truniger, Frde. 2013. Introduction: From the Aesthetic to the Dynamic. In *Filmic Mapping Landscape Landscript 2*. Edited by Fred Truniger. Berlin: Jovis, pp. 13–25.

Tschumi, Bernard. 1987. *Cinegram folie: Le Parc de la Villette*. New York: Princeton Architectural Press.

Tully, Peggy. 2013. On landscape urbanism. In *The Routledge Companion to Landscape Studies*, 1st ed. Edited by Peter Howard, Ian Thompson, Emma Waterton and Mick Atha. London and New York: Routledge Taylor and Francis Group, pp. 438–49.

Turner, Tom. 1996. *City as Landscape. A Post Post-Modern View of Design and Planning*. London and New York: Routledge Taylor and Francis Group.

von Sckell, Fridrich Ludwig. 1825. *Beiträge zur Bildenden Gartenkunst für Angehende Gartenkünstler und Gartenliebhaber*. Munich: Wernersche Verlagsgesellschaft, p. 7.

Waldheim, Charles. 1999. Aerial Representation and the Recovery of Landscape. In *Recovering landscape: Essays in Contemporary Landscape Architecture*. Edited by James Corner. New York: Princeton Architectural Press, pp. 120–39.

Waldheim, Charles. 2006. *The Landscape Urbanism Reader*. New York: Princeton Architectural Press.

Wall, Alex. 1999. Programming the Urban Surface. In *Recovering Landscape Essays in Contemporary Landscape Architecture*. Edited by James Corner. New York: Princeton Architectural Press, pp. 232–49.

Wall, Ed, and Mike Dring. 2015. Landscapes of Variance Working the Gap between Design and Nature. In *Revising Green Infrastructure Concepts Between Nature and Design*. Edited by Daniel Czechowski, Thomas Hauck and Georg Hausladen. Boca Raton: CRC Press, Taylor & Francis Group, pp. 193–206.

Weaver, Warren. 1948. Science and Complexity. *American Scientist* 36: 536–44.

Weilacher, Udo. 1996. *Between Landscape Architecture and Land Art*. Basel, Berlin and Boston: Birkhäuser.

Weilacher, Udo. 2005. Bilderwelten in Bewegung. In *Büro Kiefer. Rekombinationen*. Edited by Thies Schröder, Hanns Joosten and Michael Robinson. Stuttgart: Verlag Eugen Ulmer, pp. 7–9.

Weilacher, Udo. 2008. *Syntax of Landscape: The Landscape Architecture of Peter Latz and Partners*. Basel, Boston and Berlin: Birkhäuser.

Weilacher, Udo. 2009. Learning from Duisburg-Nord. *Topos* 69: 94–97.

Weilacher, Udo. 2014. Strukturalismus in der Landschaftsarchitektur. In *Zukunft aus Landschaft gestalten Stichworte zur Landschaftsarchitektur*. Edited by Hubertus Fischer. München: Akademische Verlagsgemeinschaft München, pp. 225–28.

Weilacher, Udo. 2018. Porosity as a Structure Principle of Urban Landscapes. In *Porous City from Metaphor to Urban Agenda*. Edited by Sophie Wolfrum, Heiner Stengel, Florian Kurbasik, Norbert Kling and Sofia Dona. Basel: Birkhäuser, pp. 230–34.

Weller, Richard. 2006. An Art of Instrumentality: Thinking Through Landscape Urbanism. In *The Landscape Urbanism Reader*. Edited by Charles Waldheim. New York: Princeton Architectural Press, pp. 69–85.

Weller, Richard. 2020. Constructing an Ecological Civilization. In *Beautiful China Reflections on Landscape Architecture in Contemporary China*. Edited by Richard J. Weller and Tatum L. Hands. Los Angeles, San Francisco, New York and Shenzhen: ORO Editions, pp. 82–89.

Weller, Richard, and Tatum Hands. 2020. Engaging with Beautiful China. In *Beautiful China Reflections on Landscape Architecture in Contemporary China*. Edited by Richard J. Weller and Tatum L. Hands. Los Angeles, San Francisco, New York and Shenzhen: ORO Editions, pp. 12–17.

Wells, Herbert George. 1902. *Anticipations of the Reaction of Mechanical and Scientific Progress: Upon Human Life and Thought*. London: Chapman & Hall, p. 40.

Wirtén, Håkan. 2022. Foreword. In *Why Cities Need Large Parks Large Parks in Large Cities*. Edited by Richard Murray. London and New York: Routledge Taylor and Francis Group, p. 8.

Wirth, Louis. 1938. Urbanism as a way of life. *The American Journal of Sociology* 44: 1–24. [CrossRef]

Wolfrum, Sophie. 2018. Urbanes Gebiet. In *Porous City from Metaphor to Urban Agenda*. Edited by Sophie Wolfrum, Heiner Stengel, Florian Kurbasik, Norbert Kling and Sofia Dona. Basel: Birkhäuser, pp. 158–61.

Wolfrum, Sophie, and Sören Schöbel. 2011. Urbanism—landscape and City M.S.c. Available online: https://www.ar.tum.de/en/studiengaenge/master/urbanism-landscape-and-city-msc/ (accessed on 1 May 2013).

Wolfrum, Sophie, and Winfried Nerdinger. 2008. *Multiple City: Stadtkonzepte 1908 bis 2008*. Berlin: Jovis.

Wolfrum, Sophie, Heiner Stengel, Florian Kurbasik, Norbert Kling, and Sofia Dona. 2018. *Porous City from Metaphor to Urban Agenda*. Basel: Birkhäuser.

Wolman, Abel. 1965. The Metabolism of Cities. *Scientific American* 213: 179–90. [CrossRef]

Worster, Donald. 1993. *The Wealth of Nature: Environmental History and the Ecological Imagination*. New York: Oxford University Press.

Wu, Liangyong. 2000. Implications of Chinese traditional human settlements concept on contemporary urban design. *World Architecture* 115: 82–85.

Wu, Liangyong. 2001. *An Introduction to Sciences of Human Settlements*. Beijing: China Architecture & Building Press.

Wu, Liangyong. 2011. *Integrated Architecture*. Beijing: Tsinghua University Press.

Yang, Rui. 2020. Beyond Beauty. In *Beautiful China Reflections on Landscape Architecture in Contemporary China*. Edited by Richard J. Weller and Tatum L. Hands. Los Angeles, San Francisco, New York and Shenzhen: ORO Editions, pp. 62–65.

Yu, Kongjian. 2012. The Big Foot Revolution. In *Designed Ecologies. The Landscape Architecture of Kongjian Yu*. Edited by William B. Saunders. Berlin: Birkhäuser Verlag GmbH, pp. 42–49.

Zhao, Jijun. 2010. A Historical Inquiry about "National Landscaping and Gardening Movement". *Chinese Landscape Architecture* 26: 56–60.

Zhu, Yufan. 2022. Nature, Ornament and Texture: The Three Facets of Complexity. *Chinese Landscape Architectue* 5: 14–24.

Zhu, Yufan, and Fanyu Meng. 2016. Potential and Strategies of Urban Open Space Transforming in BCIW. *Urban Environment Design* 2: 127–33.

Zhu, Yufan, and Yuan Xu. 2020. Toward a Space of Capability. In *Beautiful China Reflections on Landscape Architecture in Contemporary China*. Edited by Richard J. Weller and Tatum L. Hands. Los Angeles, San Francisco, New York and Shenzhen: ORO Editions, pp. 128–31.

Zöch, Peter, and Gesa Loschwitz. 2005. Riemer Park, Messestadt, Munich. *Topos* 91: 27–30.

# Index

MDPI
St. Alban-Anlage 66
4052 Basel
Switzerland
www.mdpi.com

MDPI Books Editorial Office
E-mail: books@mdpi.com
www.mdpi.com/books